Moacir Couto de Andrade Júnior
Jerusa Souza Andrade

Fermented foods in general and ethnic fermented foods in particular

AF155822

Moacir Couto de Andrade Júnior
Jerusa Souza Andrade

Fermented foods in general and ethnic fermented foods in particular

LAP LAMBERT Academic Publishing

Imprint

Any brand names and product names mentioned in this book are subject to trademark, brand or patent protection and are trademarks or registered trademarks of their respective holders. The use of brand names, product names, common names, trade names, product descriptions etc. even without a particular marking in this work is in no way to be construed to mean that such names may be regarded as unrestricted in respect of trademark and brand protection legislation and could thus be used by anyone.

Cover image: www.ingimage.com

Publisher:
LAP LAMBERT Academic Publishing
is a trademark of
Dodo Books Indian Ocean Ltd. and OmniScriptum S.R.L publishing group

120 High Road, East Finchley, London, N2 9ED, United Kingdom
Str. Armeneasca 28/1, office 1, Chisinau MD-2012, Republic of Moldova, Europe
Managing Directors: Ieva Konstantinova, Victoria Ursu
info@omniscriptum.com

Printed at: see last page
ISBN: 978-3-659-71249-4

To my grandmother (Nila) whose tasty fermented beverages (*aluá*) perfumed the June festivities and cheered the hearts of everyone.

Moacir Couto de Andrade Júnior

To my loving and caring mother Nina who has always produced, prepared and served all healthy (unfermented and fermented) foods to her family.

Jerusa Souza Andrade

CONTENTS

AUTHORS' PREFACE

At the beginning of April 2011, there was a flood of requests for original articles and reviews on the biochemistry of Amazonian fruits in the postharvest period. Our scientific research focuses on the nutritional content and the enzymology of these fruits at different ripening stages. As we first thought of the academic formation of our graduate and postgraduate students in this specific area, we believed that it would be instructive to elaborate a didactic material (a book, a monograph or an in-depth review), showing them the most apparent and practical aspect of enzyme activity, *i.e.* the food fermentation.

To ensure a clear understanding of such a vast subject in a globalized world, we decided to start from the basic knowledge regarding concepts (principles), definitions (conceptual limits of terms), evolutionary aspects, historical background, microbiological and biochemical (enzymological) bases, technological aspects as well as health benefits and harms of fermented foods.

With these fundamentals, it would be perhaps easier for the students – and even for the *grand public* too – to understand not only fermented foods (in general), but also to integrate them into those other aspects related to the ethnic fermented foods of different parts of the world, *i.e.* the pertinent socioeconomic aspects, habits, traditions and beliefs, especially of the Amazonian natives (in particular).

Thus, the title of the present book reflects exactly its conception as we described above. Many metabolic pathways are interconnected, and some steps were recurrently discussed (*i.e.* from other angles) for greater understanding. Most of the time, the footnotes and the figures brought complementary information on the subjects discussed throughout the five chapters, enriching their contents without interfering with their flow of reading. On the other hand, the tables were more integrated into the texts and helped to summarize pertinent information. We provided a thorough update of the literature on each theme. The references were numbered and listed in order of appearance in the text.

As discussed in detail in this book, fermentation may be the oldest known form of food biotechnology. Fermented foods have been widely used in human diets for thousands of years. Today, they are at the crossroads of ethnicity, culture and nutrition.

Fermented foods range from staple foods such as those prepared with starchy substrates (e.g. cassava) to delicacies. Microorganisms are effectively able to use those substrates as energy sources through their rich enzymatic equipment, transforming plain foods into more elaborated foodstuffs such as cheeses, wines, table olives and chocolates.

In addition to being very palatable, fermented foods provide energy-rich macronutrients (lipids, proteins and carbohydrates), micronutrients (minerals, water-soluble vitamins and fat-soluble vitamins), bioactive substances (mostly flavonoids), probiotics (e.g. lactic acid bacteria), variety in the diet and improvement in the organoleptic properties and digestibility of foods.

However, the predominance of macronutrients varies according to the fermented foods, e.g. carbohydrates predominate in yogurts, whereas lipids predominate in cheeses. This same statement is applicable to the bioactive substances present in diverse fermented foods and herein discussed. The high value of fermented foods places them into two categories of food: energy-dense and nutrient-dense foods.

Considering the multi-dimensional character of fermented foods as well as the entire scientific endeavor already devoted to elucidate their complexity, they drew great interest in the past (with enormous breakthroughs), but many pertinent questions still remain unanswered nowadays. The authors of this book critically discussed the past and the present of fermented foods and tried to envision their future perspectives.

Moacir Couto de Andrade Júnior, MD, M. Sc.
Jerusa Souza Andrade, Prof. Dr.

INTRODUCTION TO FERMENTED FOODS

The human genome was probably selected during the late Paleolithic era (50,000-10,000 BC) when food was essentially obtained by hunting and gathering [1, 2]. The advent of agriculture during the Holocene (10,000 BC) inscribed the first major inflection point in food yield and permanently changed the character of the human condition [2-5]. With the emergence of rudimentary fermented foods (e.g. the first vinegar, probably a result of spoiled wine), there has followed the appearance of distinctive foodstuffs, and bread has become the prominent archetypal product of cereal civilizations such as the Egyptian [6-9]. According to historical records, natural (or leavened) bread was first made in Egypt (ca. 5000 BC) [10]. In natural bread, alcoholic fermentation (by yeasts) as well as acid fermentation (by lactic acid bacteria) occurs simultaneously, giving bread its flavor and high palatability [10]. In the first century AD, the Greeks could already produce a large range of breads enriched with many ingredients such as honey, milk, sesame[*] (*Sesamum* L. spp., Pedaliaceae) and poppy seeds (*Papaver* L. spp., Papaveraceae) [7].

Thereafter, humans have ceased to be constantly famished and have progressively started to construct the concept of food preference (or food choice) in which both consumers and producers define what is good to eat [11]. As further discussed in this book, fermentation may be considered a food biotechnological legacy of the ancient world, having tremendously evolved throughout the centuries and remarkably improved the complex perception of food quality and the crucial role of food preservation. As a corollary of food preservation, the correlation between different fermented foods and food safety (or food security) has been concisely

---

[*]The scientific plant names (at least the genus, or the binomial term, always followed by the author and the family designation) were only cited once, according to the Missouri Botanical Garden (afterwards, for simplicity, only common Brazilian and English names were used). The first mention of most of the enzymes was followed by its nomenclature according to the Nomenclature Committee of the International Union of Biochemistry and Molecular Biology (NC-IUBMB). Regarding the microbiological terms, the emphasis was sometimes given only to the genus, but most of the terms were entirely cited once and abbreviated afterwards.

discussed in the **CHAPTERS 3** and **4** of this book.

Fermented foods range from staple foods such as those prepared with cereal grains and other starchy (or amylaceous) plants (e.g. wheat (*Triticum* L. spp., Poaceae), rye (*Secale cereal* L., Poaceae), corn (*Zea mays* L., Poaceae), sorghum (*Sorghum* Moench spp., Poaceae), pearl millet (*Panicum miliaceum* L., Poaceae), rice (*Oryza sativa* L., Poaceae), plantain (*Plantago* L. spp., Plantaginaceae), breadfruit (*Artocarpus* J.R. Forst. & G. Forst. spp., Moraceae), sweet potatoes – *Ipomoea batatas* (L.) Lam., Convolvulaceae), to delicacies such as cheeses, wines, table olives and chocolates [12, 13]. In addition to being very palatable, fermented foods provide energy-rich macronutrients (e.g. carbohydrates, lipids and proteins), micronutrients such as minerals (e.g. iron), water-soluble vitamins (e.g. vitamin B_{12}, folate), fat-soluble vitamins (e.g. menaquinone or vitamin K_2), functional ingredients (e.g. phenolic compounds, probiotics), variability in the diet as well as improvement in sensory properties and digestibility [14-22].

Today, fermented foods are widely spread in human alimentation. In Southeast Asia, for instance, fermented foods are prevalent and balance the fluctuation in food availability in the area during the stage of monsoonal circulation [23]. Fermented foods are at the crossroads of ethnicity, culture and nutrition and naturally attract the interest of food scientists and food enzymologists. With better understanding of fermentation enzymes, food biotechnologists will be able to develop newer and more efficient methods for the processing and storage of fermented foods [24].

Thus, the first major purpose of the present book was not only to update the literature concerning the fermented foods (in general), but also to highlight the pertinent biochemical and nutritional aspects. The thematic development always aimed at clarity, but to achieve it, some digressions were made here and there as follows.

The detailed historical background of fermented foods (in general) was developed in **CHAPTER 1**. Nevertheless, historical notes were also added where appropriate in the book. The literature is scarce regarding the historical background of fermented foods in the indigenous civilizations of the Brazilian Amazonia. Hence, it was preferentially discussed in **CHAPTER 5** reserved to this theme (in particular).

The fundamental terminology related to the fermentation processes and the fermented foods were clarified in **CHAPTER 2**.

The microbiological and biochemical (enzymological) bases of the major fermentation pathways were discussed in **CHAPTER 3** in the following order: **(3.1)** alcoholic fermentation, **(3.2)** acetic acid fermentation, **(3.3)** lactic acid fermentation, **(3.4)** butyric acid fermentation, **(3.5)** propionic acid fermentation and **(3.6)** succinic acid fermentation. A special topic on the importance of Krebs cycle in microbial metabolic pathways was developed in subsection **(3.7)**.

The related biomedical aspects were discussed in **CHAPTER 4** as follows: **(4.1)** nutritional concepts applied to fermented foods, **(4.2)** the nutritional value and health benefits of fermented foods, highlighting the clinical aspects and pathophysiological mechanisms interrelated, **(4.3)** the importance of biogenic amines in the fermented food products and other potential detrimental effects of fermented foods on human health (e.g. foodborne diseases such as listeriosis and botulism).

The socioeconomic relevance of fermented foods in the present human diets worldwide was discussed in **CHAPTER 5**.

In this last (but not least) chapter of the book, the authors attempted to clarify the importance of the cultural milieu in the emergence and the acceptance of ethnic fermented foods in the marketplaces. To fulfill this (second major) purpose, after having referred to these aspects in other parts of the book, the authors, with some more detail, referred to old techniques of food fermentation; more specifically, those of Amazonian fruits such as graviola (*Annona muricata* L., Annonaceae), camu-camu (*Myrciaria dubia* (Kunth) McVaugh, Myrtaceae Juss.), cupuaçu (*Theobroma grandiflorum* (Willd. ex Spreng.) K. Schum., Malvaceae Juss.), manioc or cassava (*Manihot esculenta* Crantz, Euphorbiaceae), pupunha or peach palm (*Bactris gasipaes* Kunth, Arecaceae), used by the natives of the Brazilian Amazonia in order to exemplify how simple operational gestures are capable of extracting tasty and nutritious ingredients and composing numerous culinary preparations (*caiçuma* and other alcoholic beverages, chocolate, *tucupi*, among others). Some of these ethnic fermented foods have expanded globally, and their commercial interest has clearly increased over the past few years.

The humanity currently faces major interrelated problems such as the disturbing demographic explosion (more than seven billion people), a profound socioeconomic (and ideological) crisis and a degrading state of chronic hunger in certain regions of the planet and, in contrast, food abundance, leading to obesity and (or) type 2 diabetes mellitus in other parts of the world.

Microorganisms may become efficient sources of macromolecules (e.g. enzymes) of great biotechnological importance as well as of bioactive molecules (e.g. vitamins) of established biomedical interest, transforming empty calorie foods (e.g. rich in sugars) in foods with high nutritional density. However, these beneficial results largely depend on the scientific dexterity with which those microorganisms are manipulated.

This book attempted to clarify, above all, the fundamental aspects of this vast theme (e.g. the joint approach of the microbiological and biochemical (enzymological) bases of the main fermentation pathways). The present-day importance of short chain fatty acid fermentation products of the gut microbiome was also discussed in **CHAPTER 4**. As a final point, the evolutionary aspects of fermented foods were discussed in due course of the book. **Figure 1**.

Figure 1:
Bread: one of the main components of human diet since ancient times.
Carl von Linné (1707-1778) and his followers praised bread not only as the core component of diet, but also for its versatile role both in health and in disease [25, 26]. The quantitative and qualitative aspects of bread consumption are further discussed in the following two references [27, 28]. Besides its energy-rich content, thiamine (or vitamin B_1), folate, health authorities have long used bread to supply faulty nutrients (e.g. iodine, vitamin A precursor or β-carotene) to deprived populations [29-31]. For instance, iodine intake from fast food items – a major source of nutrition for many Americans – may be low unless iodinated bread, milk shakes or fish are consumed [29]. There is a potential contribution of bread buns fortified with β-carotene-rich sweet potato in Central Mozambique [32].

REFERENCES

[1]   N. Halberg, M. Henriksen, N. Söderhamn, B. Stallknecht, T. Ploug, P. Schjerling, and F. Dela, "Effect of intermittent fasting and refeeding on insulin action in healthy men," *Journal of Applied Physiology*, vol. 99, no. 6, pp. 2128-2136, 2005.

[2]   R. J. Lincoln, G. A. Boxshall, and P. F. Clark, *A Dictionary of Ecology, Evolution and Systematics*, Cambridge University Press, London, UK, 1983.

[3]   F. P. Miller, "After 10,000 years of agriculture, whither agronomy?" *Agronomy Journal*, vol. 100, no. 1, pp. 22-34, 2008.

[4]   M. Jones and T. Brown, "Agricultural origins: the evidence of modern and ancient DNA," *The Holocene*, vol. 10, no. 6, pp. 769-776, 2000.

[5]   T. A. Brown, "How ancient DNA may help in understanding the origin and spread of agriculture," *Philosophical Transactions of the Royal Society B*, vol. 354, pp. 89-98, 1999.

[6]   B. Guangrun, "A study on ecological — geographical environment of the places of main origin of ancient civilizations," *Journal of Northeast Normal University*, 1991.

[7]   E. Nistor, E.-L. Sfetcu, and I.-C. Sfetcu, "Our daily bread': history... and stories," *Scientific Papers, Series A, Agronomy*, vol. LVII, pp. 432-440, 2014.

[8]   H. R. Komeili and Z. Sheikholeslami, "Replacement effect of wheat flour with barley flour and hull-less barley flour on the bread porosity and color," *Advance in Agriculture and Biology*, vol. 2, no. 1, pp. 39-43, 2014.

[9]   M. Cheryan, "Acetic acid production," in *Encyclopedia of Microbiology*, M. Schaechter, Ed., pp. 145-149, Elsevier Inc., New York, NY, USA, 3rd edition, 2009.

[10]   H.-C. Chung, B. Y. Jeong, and G. D. Han, "Optimum conditions for combined application of *Leuconostoc* sp. and *Saccharomyces* sp. to sourdough," *Food Science and Biotechnology*, vol. 20, no. 5, pp. 1373-1379, 2011.

[11]   M. L. Smith, "The archaeology of food preference," *American Anthropologist*, vol. 108, no. 3, pp. 480-493, 2006.

[12]   J. Haydersah, I. Chevallier, I. Rochette, C. Mouquet-Rivier, C. Picq, T. Marianne-Pepin, C. Icard-Vernière, and J.-P. Guyot, "Fermentation by amylolytic lactic acid bacteria and consequences for starch digestibility of plantain, breadfruit, and sweet potato flours," *Journal of Food Science*, vol. 77, no. 8, pp. M466-M472, 2012.

[13]   J. De Dea Lindner, A. L. B. Penna, I. M. Demiate, C. T. Yamaguishi, M. R. M. Prado, and J. L. Parada, "Fermented foods and human health benefits of fermented functional foods," in *Fermentation Processes Engineering in the Food Industry*, C. R. Soccol, A. Pandey, and C. Larroche, Eds., pp. 263-297, CRC Press, Boca Raton, FL, USA, 2013.

[14]   A. A. O. Ogunshe and K. O. Olasugba, "Microbial loads and incidence of food-borne indicator bacteria in most popular indigenous fermented food condiments from middle-belt and southwestern Nigeria," *African Journal of Microbiology*

Research, vol. 2, no. 12, pp. 332-339, 2008.

[15]   M. Lu, Y. Toshima, X. L. Wu, X. Zhang, and Y. Q. Cai, "Inhibitory effects of vegetable and fruit ferment liquid on tumor growth in Hepatoma-22 inoculation model," *Asia Pacific Journal of Clinical Nutrition*, vol. 16, supplement 1, pp. 443-446, 2007.

[16]   J. Chakrabartyl, G. D. Sharma, and J. P. Tamang, "Substrate utilisation in traditional fermentation technology practiced by tribes of North Cachar Hills District of Assam," *Assam University Journal of Science & Technology: Biological Sciences*, vol. 4, no. 1, pp. 66-72, 2009.

[17]   H. W. Lopez, F. Leenhardt, C. Coudray, and C. Remesy, "Minerals and phytic acid interactions: is it a real problem for human nutrition?" *International Journal of Food Science and Technology*, vol. 37, no.7, pp. 727-739, 2002.

[18]   W. Sybesma, M. Starrenburg, M. Kleerebezem, I. Mierau, W. M. de Vos, and J. Hugenholtz, "Increased production of folate by metabolic engineering of *Lactococcus lactis*," *Applied and Environmental Microbiology*, vol. 69, no. 6, pp. 3069-3076, 2003.

[19]   A. S. Adegoke, J. A. Akinyanju, and J. E. Olajide, "Fermentation of aflatoxin contaminated white dent maize (*Zea mays*)," *Research Journal of Medical Sciences*, vol. 4, no. 3, pp. 111-115, 2010.

[20]   Y. Abebe, A. Bogale, K. M. Hambidge, B. J. Stoecker, K. Bailey and R. S. Gibson, "Phytate, zinc, iron and calcium content of selected raw and prepared foods consumed in rural Sidama, Southern Ethiopia, and implications for bioavailability," *Journal of Food Composition and Analysis*, vol. 20, no. 3-4, pp. 161-168, 2007.

[21]   M. A. Rishavy and K. L. Berkner, "Vitamin K oxygenation, glutamate carboxylation, and processivity: defining the three critical facets of catalysis by the vitamin K-dependent carboxylase," *Advances in Nutrition*, vol. 3, pp. 135-148, 2012.

[22]   J. W. Suttie and S. L. Booth, "Vitamin K," *Advances in Nutrition*, vol. 2, pp. 440-441, 2011.

[23]   S. V. Law, F. Abu Bakar, D. Mat Hashim, and A. Abdul Hamid, "Popular fermented foods and beverages in Southeast Asia," *International Food Research Journal*, vol. 18, no. 2, pp. 475-484, 2011.

[24]   R. W. Owusu-Apenten, *Introduction to Food Chemistry*, CRC Press, Boca Raton, FL, USA, 2005.

[25]   L. Räsänen, "Of all foods bread is the most noble: Carl von Linné (Carl Linneaus) on bread," *Scandinavian Journal of Food and Nutrition*, vol. 51, no. 3, pp. 91-99, 2007.

[26]   P.-O. Ouellet, "Les oeuvres d'art dans les intérieurs domestiques," *Cap-aux-Diamants: La revue d'histoire du Québec*, no. 110, pp. 9-13, 2012.

[27]   J. Sesiano, "Chapter (B-XVII) on the consumption of bread by men," in *The Liber mahameleth, Sources and Studies in the History of Mathematics and Physical Sciences*, pp. 999-1005, Springer, Lausanne, Switzerland, 2014.

[28]   R. Prättälä, V. Helasoja, and H. Mykkänen, "The consumption of rye bread and white bread as dimensions of health lifestyles in Finland," *Public Health Nutrition*, vol. 4, no. 3, pp. 813-819, 2001.

[29]   S. Y. Lee, A. M. Leung, X. He, L. E. Braverman, and E. N. Pearce, "Iodine content in fast foods: comparison between two fast-food chains in the United States," *Endocrine Practice*, vol 16, no. 6, pp. 1-2, 2010.

[30]   G. Kidane, K. Abegaz, A. Mulugeta, and P. Singh, "Nutritional analysis of vitamin A enriched bread from orange flesh sweet potato and locally available wheat flours at Samre Woreda, Northern Ethiopia," *Current Research in Nutrition and Food Science*, vol. 1, no. 1, pp. 49-57, 2013.

[31]   M. G. Abubakar, A. Y. Abbas, and B. A. Faseesin, "Potassium bromate content of bread produced in Sokoto metropolis," *Nigerian Journal of Basic and Applied Sciences*, vol. 16, no. 2, pp. 183-186, 2008.

[32]   J. W. Low and van P. J. Jaarsveld, "The potential contribution of bread buns fortified with beta-carotene-rich sweet potato in Central Mozambique," *Food and*

Nutrition Bulletin, vol. 29, no. 2, pp. 98-107, 2008.

CHAPTER 1

THE HISTORICAL BACKGROUND OF FERMENTED FOODS

Since prehistoric times, various methods have been used to process and preserve foods, and fermentation may be the oldest known form of food biotechnology [1-3]. Food biotechnology is chiefly concerned with the use of food-grade microorganisms (yeasts, filamentous fungi and some bacteria) in industrial processes [4, 5]. Humankind has long made use of natural (or spontaneous) fermentation of diverse food items, which include bread, alcoholic beverages, dairy products, vegetable products and meat products [1, 6-9]. It is assumed that the art of cheese making was developed almost as soon as agriculture began in the Fertile Crescent between the Tigris and the Euphrates rivers (present-day Iraq), at the time when plants and animals were being domesticated [10, 11]. Although there is no precise record of the date for the first fermented milk, goats were initially domesticated in Mesopotamia about 5000 BC and goat milk stored warm in gourds in a hot climate naturally formed a curd (or soured milk) [12, 13].

Archaeological evidence has been found for the production of a fermented beverage in China in 7000 BC and of wine in Iran and Egypt in 6000 BC and 3000 BC, respectively [14, 15]. A fermented soybean food (soy sauce) most likely arrived in Japan from China with the introduction of Buddhism in 3000 BC [10, 16]. Originating from China, tea was first cited as early as in 100 BC in the Shen Nong's Herbal Classic – widely considered as the oldest book on oriental herbal medicine and the foundation of traditional Chinese medicine – for its detoxification effects [16]. The Chinese are claimed to be the first ones who mentioned the production of raw cured ham, but other authors of the ancients also discussed the process [17].

Moreover, it is believed that the Sumerians developed brewing in approximately 2000 BC [10, 16]. From this time, it is believed that these

15

fermentation technologies expanded from Mesopotamia into other parts of the world [12]. Consequently, the cultivation of grapevines and the production of wine spread across the Mediterranean Sea toward Greece (2000 BC), Italy (1000 BC), northern Europe (100 AD) and America (1500) [12]. In the sixteenth century, beverages produced from cocoa were essential to social and ritual occasions among the Aztecs, and cacao was one of the most valued commodities in Mesoamerica (the geographic region between northern Mexico and southern Panama) [18, 19]. In the eighteenth century (1737), Linné named the cocoa tree *Theobroma* – food of god[†] [20, 21]. The eighteenth century was particularly rich in the creation of French fermented delicacies such as the Gruyère cheese (1722), the Moët and Chandon champagne (1743), the Camembert cheese (1791) [22]. All these fermentation techniques were unquestionably artisanal in nature and there could have been no appreciation of the role of microorganisms at that time [23].

However, in the late eighteenth century, the British Industrial Revolution exerted a profound impact on western cultures, increasing demographic concentrations in urban areas and paving the way for the emergence of technological societies [23-25]. Most crucially, in the second half of the nineteenth century (the Pasteur era[‡]), the blooming of microbiology as a science resulted in the biological basis of fermentation being understood for the first time, and the essential role of bacteria, yeasts and molds in the generation of fermented foods was elucidated [23, 26, 27]. Moreover, the discovery by Eduard Buchner (1860-1917) that the juice from a cell-free yeast press retained the fermenting capacity of the living cells is considered a significant landmark in biochemical (enzymological) research, as shown in **Figure 2** [27-37]. The discovery of cell-free fermentation set the stage for modern biochemistry [38]. These major scientific facts ultimately resulted in more controlled and efficient (inoculated) fermentations, which are quicker and more reliable than

---

[†]During fermentation, aroma precursors (e.g. free amino acids, short chain peptides, reducing sugars) are formed, from which the typical, pleasant, cocoa aroma is suggested to be generated during the subsequent roasting process [20].
[‡]Louis Pasteur (1822-1895) [26].

spontaneous (or uninoculated) fermentations, and assure predictable production and quality of fermented foods [8, 23, 39].

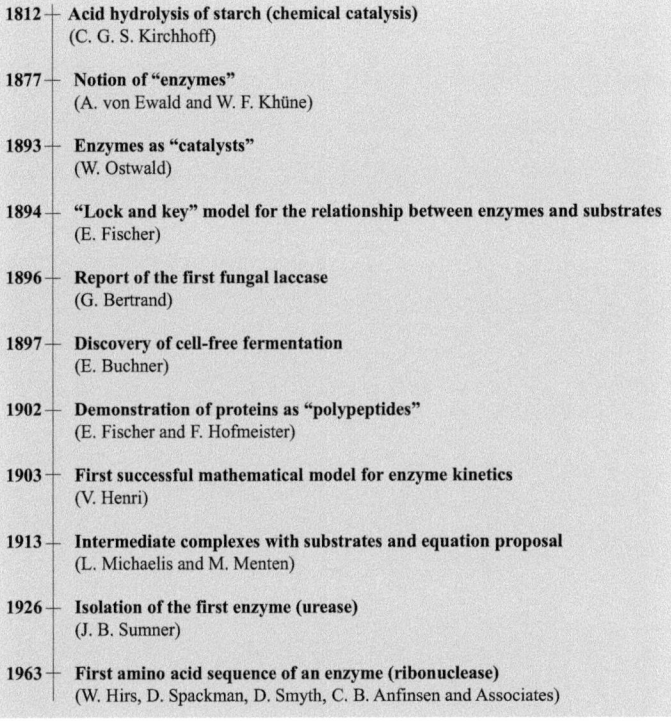

Figure 2:
Timeline of some enzymological breakthroughs (adapted from [27-37]).

The twentieth century was significantly enriched by biotechnology, defined as the multidisciplinary science applied to the manipulation of living organisms (or components of living organisms) at the molecular level, with the aim of generating useful products and improving human health, industrial efficiency and food production [40]. Enzymes then became essential biotechnological tools, especially in the pharmaceutical and food industries. The first microbial enzyme used in the wine industry was a commercial pectinase (generic term) from the fungus of the genus *Aspergillus*, which contained varying amounts of pectinesterase (EC 3.1.1.11),

17

polygalacturonase (EC 3.2.1.15), pectin lyase (EC 4.2.2.10) and hemicellulase (EC 3.1.1.73) in small amounts [41]. The use of these enzymes gives the winemaker many advantages, among which the clarification (turbidity loss) of the beverage [42-44].

Subsequently, commercial microbial enzymes have been selected for their high activities, versatility on diverse substrates and durability for industrial uses [45]. Bacteria belonging to the genus *Bacillus* are undoubtedly the most important sources of several commercial exogenous enzymes such as α-amylase (EC 3.2.1.1) for syrup production, baking and brewing, and pullulanase (EC 3.2.1.41) for starch (amylopectin) degradation [15, 46-48]. These and other examples are illustrated in **Figure 3**.

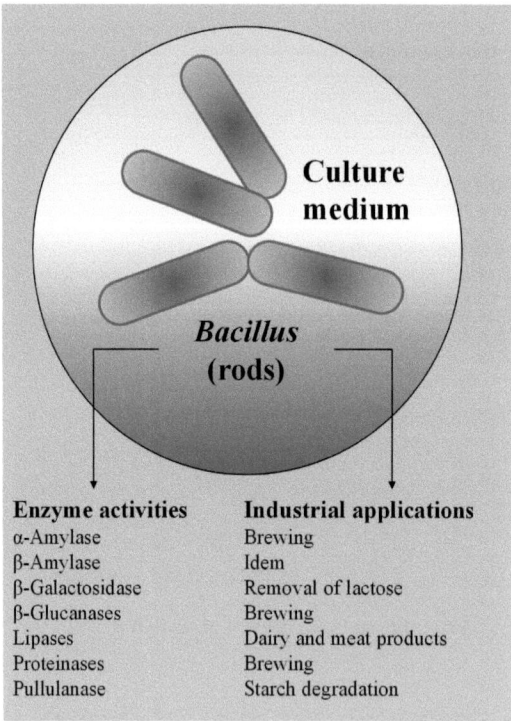

Figure 3:
Some examples of *Bacillus* enzymes applied in the food industry (adapted from [15]).

18

At the dawn of the twenty-first century, biotechnology presented the humanity with the genome sequences of many species, ranging from prokaryotes such as the Gram-negative bacteria *Haemophilus influenzae* in 1995 and *Escherichia coli* in 1997, to eukaryotes such as the baker's yeast *Saccharomyces cerevisiae* in 1996 and the molecular geneticists' model plant *Arabidopsis thaliana* (L.) Heynh. (Brassicaceae) in 2000, including the *Homo sapiens* in 2001 [22, 49-53]. As expected, these scientific breakthroughs have revealed not only a great deal of fundamental information concerning microbial metabolism and research (productive strains), plant physiology and human pathology, but they have also shaped innovative scientific perspectives on food biotechnology and many other areas (e.g. bioenergetics) for the new millennium.

Nevertheless, the need for continued growth of alternative energy sources has resulted in complex conflicts of interest in the use of natural resources (e.g. sugar cane – *Saccharum officinarum* L., Poaceae –, corn), usually employed in food manufacture, instead of the production of biofuels (e.g. bioethanol). Schematically, the first-generation bioethanol has a head start of more than half a century and is produced from the substrates abovementioned (the USA and Brazil are the two largest producers from corn and sugar cane, respectively) [54-56]. The second-generation bioethanol is also denoted as cellulosic ethanol, as it is mainly produced from agricultural residues (or other lignocellulosic raw materials) [56]. Its major obstacle is the accessibility of cellulose affected by hemicelluloses, acting as physical barriers to enzymatic saccharification (*i.e.* the conversion of cellulose into glucose and that of hemicelluloses into pentoses and hexoses) [57, 58]. The third-generation bioethanol represents fuel ethanol produced from algal biomass [59]. The fourth-generation bioethanol is the one in which the plant itself (genetic modification of a characterized variety) will be genetically able to produce the necessary enzymes that participate in the process of digestion of its own cell wall [60]. The impact of fossil fuel on the environment, especially the global warming and the harmful effects of carbon emissions, has created a new demand for clean and sustainable energy sources [61]. The multidisciplinary engagement in this area is essential to appease the aforementioned conflicts (food versus fuels) and to achieve the specific goals. This joint endeavor holds great promises for the future of humanity.

1.1 Special topic on popular fermented vegetables

At present, certain species of (table) olives (e.g. *Olea europaea* var. *communis* Sol., Oleaceae), cucumbers (e.g. *Cucumis sativus* L., Cucurbitaceae) as pickles and cabbage (e.g. *Brassica oleracea* var. *capitata* L., Brassicaceae) as sauerkraut (or kraut), among few other examples, account for the majority of the vegetables commercially fermented [62, 63] – **Figure 4**.

Table olives

Pickles

Sauerkraut (or kraut)

Figure 4:
Fermented vegetables: examples among the most popular in the world.

1.1.1 Table olives

Although early exploitation and use of wild olive trees (*i.e.* oleasters) have been documented since the Neolithic period from the Near East to Spain, it is usually

20

accepted that the domestication of the olive tree – characterized by vegetative propagation of the best cultivated genotypes that may have preceded orchard establishment – began in the Near East approximately 6000 years ago [63-65]. Olive is a slow-growing, long-lived, evergreen tree and is a suitable model to study the origin of cultivars due to its long lifespan, resulting in the existence of both centennial and millennial trees across the Mediterranean Basin [22, 64, 66]. Despite this opulent chronology fit for scientific investigation, the historical knowledge about olive fermentation remains poor. It was only mentioned in the first century by Lucius Columela in De Re Rustica (42 BC), which is the first reference on how to prepare table olives [67].

The current practice of olive fermentation has become a large-scale production worldwide, and the world production of table olives surpasses 2.2 million tons per year (the Mediterranean countries are the main producers) [62]. However, more than 90% of the total worldwide olive production is used for oil and only 7% to 10% are consumed as table olives [22].

The olive fruit is a drupe that contains a bitter phenolic component (oleuropein[§]), a low-sugar concentration (usually between 2.5-6.0%) and a high-fat content (between 9-30%), although these values can differ with maturity degree and olive variety [67-70].

The spontaneous fermentation of table olives mainly depends on lactic acid bacteria and, in particular, on *Lactobacillus plantarum*, which plays an important role in the degradation of oleuropein via β-glucosidase and esterase activities [71]. **Figure 5.**

[§]Otherwise, it displays a broad variety of health beneficial properties (e.g. antioxidant activity) [69, 70].

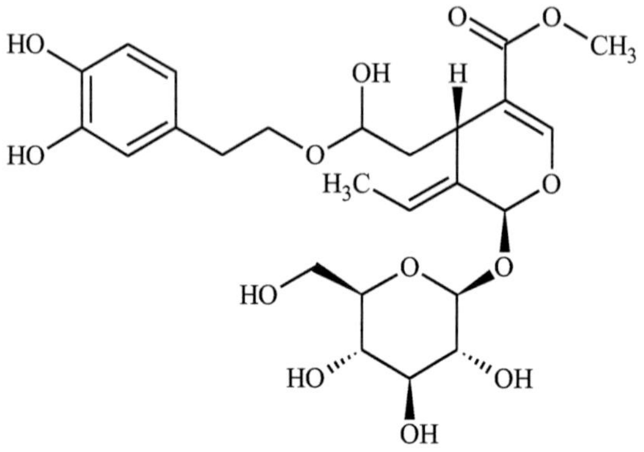

Figure 5:
Oleuropein: named and first studied by E. Bourquelot and J. Vintilesco at the beginning of the twentieth century [72]. Its enzymatic degradation (and consequent olive debittering) during bacterial lactic acid fermentation gives the typical acidic flavor of table olives [69, 70]. Additionally, this phenolic compound may be incorporated into other fermented foods (e.g. yogurts) as a supplementary functional ingredient without undesirable sensory changes e.g. in flavor [69].

1.1.2 Pickles

The combination of organic acids (e.g. lactic acid, acetic acid), spices and sugar with cucumbers (or other vegetables, e.g. eggplants – *Solanum melongena* L., Solanaceae) creates the acidic food product known as pickles [62, 73].

It is believed that cucumbers were first fermented around 2000 BC in the Middle East and early written records of cucumber pickles come from paper fragment remains of a play (*The Taxiarchs*) by the Greek writer Eupolis (429-412 BC) [74].

Today, cucumber pickles are made in Africa, Asia, United States and Latin America [62]. For instance, in the United States, companies producing cucumber pickles can have at one location as many as 1000 fermentation tanks of 40,000-liter capacity, totaling 40 million liters [74]. Lactic acid bacteria such as *Lactobacillus brevis*, *Leuconostoc mesenteroides* and *Pediococcus pentosaceus* are involved in the pickling process [22, 62, 75].

1.1.3 Sauerkraut (or kraut)

Sauerkraut (or kraut) is a product that results from the lactic acid fermentation of fresh shredded, salted, white cabbage [62]. Sauerkraut was historically an essential food for naval forces and seafarers who had little access to fresh vegetables and fruits that normally would have served as a source of vitamin C to prevent scurvy [62].

Many details (microbiological, biochemical) of the sauerkraut fermentation were described as long ago as the 1930s, and this interest has undoubtedly been due, in large part, to the very nature of the fermentation process, in that it involves several different naturally occurring microorganisms acting as part of a complex ecosystem [22]. The lactic acid bacteria abovementioned, including *Leuconostoc fallax* and other strains, are preeminent in this fermentation process [22, 76, 77].

It has been reported that the fermentation of cabbage enhances the formation of a potent chemopreventive agent, ascorbigen, a compound that results from the reaction of indole-3-carbinol or I3C (derived from the most commonly studied glucosinolate, glucobrassicin) and vitamin C in sauerkraut [76-78].

In other words and more precisely, glucosinolates are sulfur-containing glucosides, which are classified into three different classes: aliphatic glucosinolates, indole glucosinolates and aromatic glucosinolates [78]. During shredding, the glucosinolate glucobrassicin is transformed into I3C by the action of myrosinase (β-thioglucoside glucohydrolase; EC 3.2.3.1.) and during fermentation, as the pH decreases, the I3C reacts nonenzymatically with L-ascorbic acid to yield ascorbigen [77, 78].

According to recent data, ascorbigen showed much stronger antioxidant activity than did ascorbic acid and Trolox (a hydrophilic analogue of vitamin E and a free radical scavenger) [79, 80]. **Figure 6.**

23

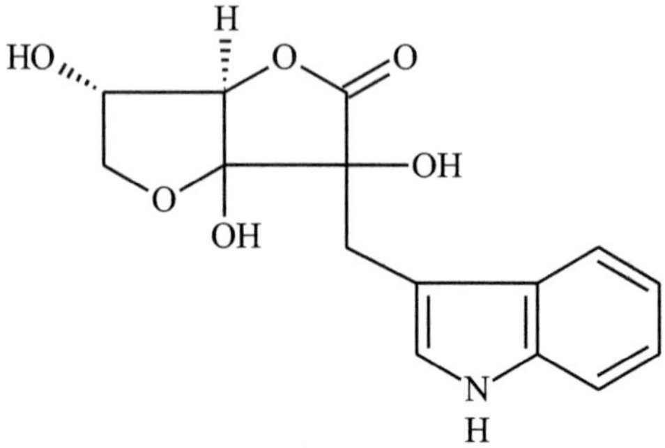

Figure 6:
Ascorbigen: the term was proposed by J. C. Pal and B. C. Guha in the 1930s for the ascorbic acid held in the "combined form" in plant tissues [81]. It has been experimentally demonstrated that dietary indoles (e.g. I3C and ascorbigen) increase the activity of phase I and phase II xenobiotic metabolic enzymes in the liver and intestinal mucosa [82]. These enzymes were reviewed in the following reference [83].

REFERENCES

[1]   S. V. Law, F. Abu Bakar, D. Mat Hashim, and A. Abdul Hamid, "Popular fermented foods and beverages in Southeast Asia," *International Food Research Journal*, vol. 18, no. 2, pp. 475-484, 2011.

[2]   S. Parveen and F. Hafiz, "Fermented cereal from indigenous raw materials," *Pakistan Journal of Nutrition*, vol. 2, no. 5, pp. 289-291, 2003.

[3]   M. E. Guerzoni, "Human food chain and microorganisms: a case of co-evolution," *Frontiers in Microbiology*, vol. 1, no. 106, pp. 1-2, 2010.

[4]   A. Querol, C. Belloch, M. T. Fernández-Espinar, and E. Barrio, "Molecular evolution in yeast of biotechnological interest," *International Microbiology*, vol. 6, no. 3, pp. 201-205, 2003.

[5]   J. E. Pérez-Ortín, J. García-Martínez, and T. M. Alberola, "DNA chips for yeast biotechnology. The case of wine yeasts," *Journal of Biotechnology*, vol. 98, no. 2-3, pp. 227-241, 2002.

[6]   M. Keller, "A historical overview of alcohol and alcoholism," *Cancer Research*, vol. 39, pp. 2822-2829, 1979.

[7]   S. Jube and D. Borthakur, "Recent advances in food biotechnology research," in *Food Biochemistry and Food Processing*, Y. H. Hui, Ed., pp. 35-69, Blackwell Publishing, Iowa, USA, 2006.

[8]   C. Varela, T. Siebert, D. Cozzolino, L. Rose, H. McLean, and P.A. Henschke, "Discovering a chemical basis for differentiating wines made by fermentation with 'wild' indigenous and inoculated yeasts: role of yeast volatile compounds," *Australian Journal of Grape and Wine Research*, vol. 15, no. 3, pp. 238-248, 2009.

[9]   C. Boulton and D. Quain, *Brewing Yeast and Fermentation*, Blackwell Science, Oxford, UK, 2001.

[10]   R. P. Ross, S. Morgan, and C. Hill, "Preservation and fermentation: past, present and future," *International Journal of Food Microbiology*, vol. 79, no. 1-2, pp. 3-16, 2002.

[11]   P. G. Bille, P. Hiwelepo, and E. L. Keya, "Examining the need for the use of calcium in the processing of Gouda cheese made from pasteurized milk," *The Journal of Food Technology in Africa*, vol. 6, no. 2, pp. 44-47, 2001.

[12]   J. L. Legras, D. Merdinoglu, J. M. Cornuet, and F. Karst, "Bread, beer and wine: *Saccharomyces cerevisiae* diversity reflects human history," *Molecular Ecology*, vol. 16, no. 10, pp. 2091-2102, 2007.

[13]   N. Hajela, G. B, Nair, N. K, Ganguly, "Are probiotics a feasible intervention for prevention of diarrhoea in the developing world?," *Gut Pathogens*, vol. 2, no. 10, 2010.

[14]   P. Wanakhachornkrai and S. Lertsiri, "Comparison of determination method for volatile compounds in Thai soy sauce," *Food Chemistry*, vol. 83, no. 4, pp. 619-629, 2003.

[15]   C. W. Bamforth, *Food, Fermentation and Micro-organisms*, Blackwell Publishing, Ames, IA, USA 2005.

[16]   A. Vaughan, T. O'Sullivan, and D. Sinderen, "Enhancing the microbiological stability of malt and beer – a review," *Journal of the Institute of Brewing*, vol. 111,

no. 4, pp. 355-371, 2005.

[17] P. Zeuthen, "A historical perspective of meat fermentation," in *Handbook of Fermented Meat and Poultry*, F. Toldrá, Ed., pp. 3-8, Blackwell Publishing, Ames, IA, USA, 2007.

[18] J. S. Henderson, R. A. Joyce, G. R. Hall, W. J. Hurst, and P. E. McGovern, "Chemical and archaeological evidence for the earliest cacao beverages," *Proceedings of the National Academy of Sciences*, vol. 104, no. 48, pp. 18937-18940, 2007.

[19] J. Arroyo-Cabrales, O. J. Polaco, C. Laurito, E. Johnson, M. T. Alberdi, and A. L. V. Zamora, "The proboscideans (Mammalia) from Mesoamerica," *Quaternary International*, vol. 169-170, pp. 17-23, 2007.

[20] F. Frauendorfer and P. Schieberle, "Changes in key aroma compounds of criollo cocoa beans during roasting," *Journal of Agricultural and Food Chemistry*, vol. 56, pp. 10244-10251, no. 21, 2008.

[21] M. Rusconi and A. Conti, "*Theobroma cacao* L., the food of the gods: a scientific approach beyond myths and claims," *Pharmacological Research*, vol. 61, no. 1, pp. 5-13, 2010.

[22] R. W. Hutkins, *Microbiology and Technology of Fermented Foods*, Blackwell Publishing, Ames, IA, USA, 2006.

[23] E. Caplice and G. F. Fitzgerald, "Food fermentations: role of microorganisms in food production and preservation," *International Journal of Food Microbiology*, vol. 50, no. 1-2, pp. 131-149, 1999.

[24] S. B. Eaton, M. Konner, and M. Shostak, "Stone agers in the fast lane: chronic degenerative diseases in evolutionary perspective," *The American Journal of Medicine*, vol. 84, no. 4, pp. 739-749, 1988.

[25] G. D. Snooks, "Uncovering the laws of global history," *Social Evolution & History*, vol. 1, no. 1, pp. 25-53, 2002.

[26] M. H. Dominiczak, "Louis Pasteur in his laboratory: entry of chemistry into

medicine," *Clinical Chemistry*, vol. 57, no. 2, pp. 356-358, 2011.

[27]   E. B. Chain, "Landmarks and perspectives in biochemical research," *British Medical Journal*, vol. 1, no. 5429, pp. 209-220, 1965.

[28]   E. B. Chain, "Contributions from chemical microbiology to therapeutic medicine," *Proceedings of the Royal Society of Medicine*, vol. 58, no. 2, pp. 85-96, 1965.

[29]   M. Fontecave, "Understanding life as molecules: reductionism versus vitalism," *Angewandte Chemie International Edition*, vol. 49, no. 24, pp. 4016-4019, 2010.

[30]   J. Büttner, "Evolution of clinical enzymology," *Journal of Clinical Chemistry & Clinical Biochemistry*, vol. 19, no. 8, pp. 529-538, 1981.

[31]   R. A. Copeland, *Enzymes: A Practical Introduction to Structure, Mechanism, and Data Analysis*, Wiley-VCH, New York, NY, USA, 2000.

[32]   H. R. Horton, L. A. Moran, K. G. Scrimgeour, M. D. Perry, and J. D. Rawn, *Principles of Biochemistry*, Pearson Prentice Hall, NJ, USA, 2006.

[33]   K. Tipton and S. Boyce, "History of the enzyme nomenclature system," *Bioinformatics*, vol. 16, no. 1, pp. 34-40, 2000.

[34]   D. Walther, M. Ruben, and S. Rau, "Carbon dioxide and metal centres: from reactions inspired by nature to reactions in compressed carbon dioxide as solvent," *Coordination Chemistry Reviews*, vol. 182, no. 1, pp. 67-100, 1999.

[35]   G. S. Nyanhongo, G. Gübitz, P. Sukyai, C. Leitner, D. Haltrich, and R. Ludwig, "Oxidoreductases from *Trametes* spp. in biotechnology: a wealth of catalytic activity," *Food Technology and Biotechnology*, vol. 45, no. 3, pp. 250-268, 2007.

[36]   "William H. Stein – Nobel Lecture." Nobelprize.org. 25 Apr 2011, http://nobelprize.org/nobel_prizes/chemistry/laureates/1972/stein-lecture.html.

[37]   L. A. Hazelwood, J.-M. Daran, A. J. A. van Maris, J. T. Pronk, and J. R. Dickinson, "The Ehrlich pathway for fusel alcohol production: a century of research on *Saccharomyces cerevisiae* metabolism," *Applied and Environmental*

Microbiology, vol. 74, no. 8, pp. 2259-2266, 2008.

[38]   A. Kornberg, "Centenary of the birth of modern biochemistry," *The Journal of the Federation of American Societies for Experimental Biology*, vol. 11, pp. 1209-1214, 1997.

[39]   P. Raspor, F. Cus, K. P. Jemec, T. Zagorc, N. Cadez, and J. Nemanic, "Yeast population dynamics in spontaneous and inoculated alcoholic fermentations of Zametovka must," *Food Technology and Biotechnology*, vol. 40, no. 2, pp. 95-102, 2002.

[40]   P.-S. Juo, *Concise Dictionary of Biomedicine and Molecular Biology*, CRC Press, Boca Raton, FL, USA, 2002.

[41]   M. K. Bhat, "Cellulases and related enzymes in biotechnology," *Biotechnology Advances*, vol. 18, no.5, pp. 355-383, 2000.

[42]   K. Mojsov, "Use of enzymes in wine making: a review," *International Journal of Management Technology*, vol. 3, no. 9, pp. 112-127, 2013.

[43]   A. R. Tapre and R. K. Jain, "Pectinases: enzymes for fruit processing industry," *International Food Research Journal*, vol. 21, no. 2, pp. 447-453, 2014.

[44]   K. Prathyusha and V. Suneetha, "Bacterial pectinases and their potent biotechnological application in fruit processing/juice production industry: a review," *Journal of Phytology*, vol. 3, no. 6, pp. 16-19, 2011.

[45]   F. Y. Bih, S. S. H. Wu, C. Ratnayake, L. L. Walling, E. A. Nothnagel, and A. H. C. Huang, "The predominant protein on the surface of maize pollen is an endoxylanase synthesized by a *Tapetum* mRNA with a long 5′ leader," *The Journal of Biological Chemistry*, vol. 274, no. 32, pp. 22884-22894, 1999.

[46]   C. R. Silva, A. B. Delatorre, and M. L. L. Martins, "Effect of the culture conditions on the production of an extracellular protease by thermophilic *Bacillus* sp. and some properties of the enzymatic activity," *Brazilian Journal of Microbiology*, vol. 38, no. 2, pp. 253-258, 2007.

[47]   I. Rasooli, S. D. A. Astaneh, H. Borna, and K. A. Barchini, "A thermostable α-

amylase producing natural variant of *Bacillus* spp. isolated from soil in Iran," *American Journal of Agricultural and Biological Sciences*, vol. 3, no. 3, pp. 591-596, 2008.

[48]   M. Schallmey, A. Singh, and O. P. Ward, "Developments in the use of *Bacillus* species for industrial production," *Canadian Journal of Microbiology*, vol. 50, no. 1, pp. 1-17, 2004.

[49]   D. Abdulrehman, P. T. Monteiro, M. C. Teixeira, N. P. Mira, A. B. Lourenço, S. C. dos Santos, T. R. Cabrito, A. P. Francisco, S. C. Madeira, R. S. Aires, A. L. Oliveira, I. Sá-Correia, and A. T. Freitas, "YEASTRACT: providing a programmatic access to curated transcriptional regulatory associations in *Saccharomyces cerevisiae* through a web services interface," *Nucleic Acids Research*, vol. 39, supplement 1, pp. D136-D140, 2011.

[50]   S. B. Primrose and R. M. Twyman, *Principles of Gene Manipulation and Genomics*, Blackwell, Cornwall, Australia, 2006.

[51]   P. Zimmermann, M. Hirsch-Hoffmann, L. Hennig, and W. Gruissem, "GENEVESTIGATOR. Arabidopsis microarray database and analysis toolbox," *Plant Physiology*, vol. 136, no. 1, pp. 2621-2632, 2004.

[52]   K. Mochida and K. Shinozaki, "Genomics and bioinformatics resources for crop improvement," *Plant Cell Physiology*, vol. 51, no. 4, pp. 497-523, 2010.

[53]   J. C. D. Hinton, "The *Escherichia coli* genome sequence: the end of an era or the start of the FUN," *Molecular Microbiology*, vol. 26, no. 3, pp. 417-422, 1997.

[54]   D. Graham-Rowe, "Beyond food versus fuel," *Nature*, vol. 474, pp. S6-S8, 2011.

[55]   D. Ghosh and P. C. Hallenbeck, "Advanced bioethanol production," in *Microbial Technologies in Advanced Biofuels Production*, P. C. Hallenbeck, Ed., pp. 165-181, Springer, New York, NY, USA, 2012.

[56]   N. Bion, D. Duprez, and F. Epron, "Design of nanocatalysts for green hydrogen production from bioethanol," *ChemSusChem*, vol. 5, no. 1, pp. 76-84, 2012.

[57] H. Ravalason, S. Grisel, D. Chevret, A. Favel, J.-G. Berrin, J.-C. Sigoillot, and I. Herpoël-Gimbert, *"Fusarium verticillioides* secretome as a source of auxiliary enzymes to enhance saccharification of wheat straw," *Bioresource Technology*, vol. 114, pp. 589-596, 2012.

[58] S. Phuengjayaem, A. Poonsrisawat, A. Petsom, and S. Teeradakorn, "Optimization of saccharification conditions of acid-pretreated sweet sorghum straw using response surface methodology," *Journal of Agricultural Science*, vol. 6, no. 9, pp. 120-133, 2014.

[59] C. S. Goh and K. T. Lee, "A visionary and conceptual macroalgae-based third-generation bioethanol (TGB) biorefinery in Sabah, Malaysia as an underlay for renewable and sustainable development," *Renewable and Sustainable Energy Reviews*, vol. 14, no. 2, pp. 842-848, 2010.

[60] C. R. Soccol, L. P. de S. Vandenberghe, A. B. P. Medeiros, S. G. Karp, M. Buckeridge, L. P. Ramos, A. P. Pitarelo, V. Ferreira-Leitão, L. M. F. Gottschalk, M. A. Ferrara, E. P. da S. Bom, L. M. P. de Moraes, J. de A. Araújo, and F. A. G. Torres, "Bioethanol from lignocelluloses: status and perspectives in Brazil," *Bioresource Technology*, vol. 101, no. 13, 4820-4825, 2010.

[61] D. Ch. Das, A. K. Roy, and N. Sinha, "Genetic algorithm based PI controller for frequency control of an autonomous hybrid generation system," *Proceedings of the International MultiConference of Engineers and Computer Scientists*, vol. II, 2011.

[62] J. De D. Lindner, A. L. B. Penna, I. M. Demiate, C. T. Yamaguishi, M. R. M. Prado, and J. L. Parada, "Fermented foods and human health benefits of fermented functional foods," in *Fermentation Processes Engineering in the Food Industry*, C. R. Soccol, A. Pandey, and C. Larroche, Eds., pp. 263-297, CRC Press, Boca Raton, FL, USA, 2013.

[63] P. Vossen, "Olive oil: history, production, and characteristics of the world's classic oils," *HortScience*, vol. 42, no. 5, pp. 1093-1100, 2007.

[64]   J. Janick, "The Origins of Fruits, Fruit Growing, and Fruit Breeding," *Plant Breeding Reviews*, vol. 25, pp. 255-321, 2005.

[65]   G. Besnard, B. Khadari, M. Navascués, M. Fernández-Mazuecos, A. El Bakkali, N. Arrigo, D. Baali-Cherif, V. B.-B. de Caraffa, S. Santoni, P. Vargas, and V. Savolainen, "The complex history of the olive tree: from Late Quaternary diversification of Mediterranean lineages to primary domestication in the northern Levant," *Proceedings of the Royal Society B*, vol. 280, pp. 1-7, 2013.

[66]   C. M. Díez, I. Trujillo, E. Barrio, A. Belaj, D. Barranco, and L. Rallo, "Centennial olive trees as a reservoir of genetic diversity," *Annals of Botany*, vol. 108, no. 5, pp. 797-807, 2011.

[67]   F. N. Arroyo-López, J. Bautista-Gallego, F. Rodríguez-Gómez, and A. Garrido-Fernández, "Predictive microbiology and table olives," in *Current Research, Technology and Education Topics in Applied Microbiology and Microbial Biotechnology*, A. Méndez-Vilas, Ed., pp. 1452-1461, FORMATEX, Badajoz, Spain, 2010.

[68]   M. Sarbishegi, F. Mehraein, and M. Soleimani, "Antioxidant role of oleuropein on midbrain and dopaminergic neurons of substantia nigra in aged rats," *Iranian Biomedical Journal*, vol. 18, no. 1, pp. 16-22, 2014.

[69]   E. Zoidou, P. Magiatis, E. Melliou, M. Constantinou, S. Haroutounian, and A.-L. Skaltsounis, "Oleuropein as a bioactive constituent added in milk and yogurt," *Food Chemistry*, vol. 158, pp. 319-324, 2014.

[70]   N. Ghabbour, Z. Lamzira, P. Thonart, P. Cidalia, M. Markaoui, and A. Asehraou, "Selection of oleuropein-degrading lactic acid bacteria strains isolated from fermenting Moroccan green olives," *Grasas y Aceites*, vol. 62, no. 1, pp. 84-89, 2011.

[71]   M. Zago, B. Lanza, L. Rossetti, I. Muzzalupo, D. Carminati, and G. Giraffa, "Selection of *Lactobacillus plantarum* strains to use as starters in fermented table olives: oleuropeinase activity and phage sensitivity," *Food Microbiology*, vol. 34, no.

1, pp. 81-87, 2013.

[72] W. M. Walter, Jr., H. P. Fleming, and J. L. Etchells, "Preparation of antimicrobial compounds by hydrolysis of oleuropein from green olives," *Applied Microbiology*, vol. 26, no. 5, pp. 773-776, 1973.

[73] H. Shinagawa, R. Nishiyama, S. Miyao, and M. Kozaki, "Organic acid composition and quality of Japanese 'Shibazuke' pickles," *Food Science and Technology International Tokyo*, vo. 3, no. 2, pp. 170-172, 1997.

[74] F. Breidt, R. F. McFeeters, I. Perez-Diaz, and C.-H. Lee, "Fermented vegetables," in *Food Microbiology: Fundamentals and Frontiers*, M. P. Doyle and R. L. Buchanan, Eds., pp. 841-855, ASM Press, Washington, D.C., 4th edition, 2013.

[75] E. Bartkienė, L. Šernienė, G. Juodeikienė, E. Drungilas, and Ž. Valatkevičienė, "The safety, technology and sensory aspects of pasteurized and raw milk treated by solid-state fermented grain extrudates inocculated with certain lactobacilli," *Veterinarija ir Zootechnika*, vol. 65, no. 87, pp. 3-10, 2014.

[76] E. Peñas, J. M. Pihlava, C. Vidal-Valverde, and J. Frias, "Influence of fermentation conditions of *Brassica oleracea* L. var. *capitata* on the volatile glucosinolate hydrolysis compounds of sauerkrauts," *LWT - Food Science and Technology*, vol. 48, no.1, pp. 16-23, 2012.

[77] V. Krajka-Kuźniak, H. Szaefer, A. Bartoszek, and W. Baer-Dubowska, "Modulation of rat hepatic and kidney phase II enzymes by cabbage juices: comparison with the effects of indole-3-carbinol and phenethyl isothiocyanate," *British Journal of Nutrition*, vol. 105, pp. 816-826, 2011.

[78] C. Martinez-Villaluenga, E. Peñas, J. Frias, E. Ciska, J. Honke, M. K. Piskula, H. Kozlowska, and C. Vidal-Valverde, "Influence of fermentation conditions on glucosinolates, ascorbigen, and ascorbic acid content in white cabbage (*Brassica oleracea* var. *capitata* cv. Taler) cultivated in different seasons," *Journal of Food Science*, vol. 74, no. 1, pp. C62-C67, 2009.

[79] A. Tai, K. Fukunaga, A. Ohno, and H. Ito, "Antioxidative properties of

ascorbigen in using multiple antioxidant assays," *Bioscience, Biotechnology, and Biochemistry*, vol. 78, no. 10, pp. 1723-1730, 2014.

[80] D. S. Fagundes, S. Gonzalo, L. Grasa, M. Castro, M. P. Arruebo, M. A. Plaza, M. D. Murillo, "Trolox reduces the effect of ethanol on acetylcholine-induced contractions and oxidative stress in the isolated rabbit duodenum," *Revista Espanola de Enfermedades Digestivas*, vol. 103, no. 8, pp. 396-401, 2011.

[81] G. H. Carroll, "The rôle of ascorbic acid in plant nutrition," *The Botanical Review*, vol. 9, no. 1, 41-48, 1943.

[82] M. Chauhan, "A pilot study on wheat grass juice for its phytochemical, nutritional and therapeutic potential on chronic diseases," *International Journal of Chemical Studies*, vol. 2, no. 4, pp. 27-34, 2014.

[83] M. C. de Andrade Jr. and J. S. Andrade, "Amazonian fruits: an overview of nutrients, calories and use in metabolic disorders," *Food and Nutrition Sciences*, vol. 5, no. 17, pp. 1692-1703, 2014.

CHAPTER 2

THE FUNDAMENTAL TERMINOLOGY RELATED TO FERMENTATION PROCESSES, ENZYMES AND FERMENTED FOODS

2.1 The traditional fermentation processes

The etymon the closest to the current English term fermentation came from the French word *ferment*, which was derived from the Latin word *fermentum* for *fervimentum* (leaven, yeast) from *fervere* (to be boiling hot, ferment) [1-5]. The earliest uses of the term fermentation were certainly intuitive and most likely related to the action of yeast plants on sugar (with the formation of carbon dioxide and alcohol) as well as to the foaming that occurs during the manufacture of wine and beer [4, 6, 7].

With regard to tea, the most consumed beverage after water, the historical choice of the term fermentation remains controversial in the literature, in part because of the different applications of this method, *i.e.* the green tea being unfermented, the *oolong* tea[**] being partially fermented and the black tea being fully fermented [8-11]. With regard to coffee, Brazil is the largest producer in the world [12]. It can be processed by two different methods, referred to as the "wet" or the "dry" method [13, 14]. The wet method is used for coffee (e.g. *Coffea arabica* L., Rubiaceae), involving removal of the outer skin and part of the mucilage by machines, and the remaining mucilage is then removed by fermentation in water for 24-48 hours, followed by drying [13, 14]. The dry method is mainly used for coffee (e.g. *Coffea robusta* L. Linden, Rubiaceae Juss.) which has a thin pulp that allows direct drying [13, 14].

However, in a technical standpoint, the term fermentation refers to the process

---

[**]The term *oolong* is a variation of a Chinese word that means "black dragon" [8].

34

of intended chemical change in foods catalyzed by microbial enzymes [15]. Fermentation processes are usually defined and classified according to the main fermentation end products (*i.e.* the organic compounds already mentioned in the introduction of this book and further highlighted in **CHAPTER 3**) [16-20]. Nonetheless, before, it is elucidating to give some general perspective on these compounds.

The succinic acid was first purified from amber by Georgius Agricola in 1546, and it has been produced by microbial fermentation for the use in agricultural, food and pharmaceutical industries [21, 22]. It can be used as a precursor of many industrially important chemicals, including adipic acid, 1,4-butanediol, tetrahydrofuran, *N*-methyl pyrrolidinone, 2-pyrrolidinone, succinate salts and gamma-butyrolactone [21, 22].

The acetic acid was first purified in 1700 [23].

The lactic acid was first discovered in 1780 by the Swedish chemist, Carl Wilhelm Scheele, who isolated it from sour milk as impure brown syrup and gave it a name based on its linguistic origins (*Mjölksyra*) [24].

The ethyl alcohol (ethanol) was first isolated in the pure form in 1820 by Jean Dumas, who had also noticed the clinical effect of chronic alcoholism [25].

The term propionic acid, derived from the transliterated Greek words, *protos* (first) and *pion* (fat), was first described in 1844 by Johann Gottlieb, who found it among the degradation products of sugar [26].

On the other hand, it is imperative to recall that the fermentation processes can be classified into many other types, according to multiple determining factors (**Table 1**).

Table 1:
Different types of fermentation based on multiple factors (adapted from [18, 27-29]).

Determining factors	Fermentation types	Definitions and/or examples (e.g.)
Mode of cultivation	1. Batch culture (or closed culture)	Cells are cultured in a volume of liquid medium
	2. Continuous culture (or open culture)	Specific microbial growth rate relative to its theoretical maximum controlled by the external substrate concentration of the limiting nutrient
	3. Fed-batch culture	Intermediate system
Water activity	4. Submerged fermentation	Water taken as fixed amount in a static culture or under stirring conditions
	5. Solid-state fermentation	Solid substrate, restricted moisture
	6. Immobilized systems	E.g. physical adsorption, covalent bonding, entrapment, encapsulation and cross-linking
Oxygen requirements	7. Aerobic fermentation	Critical dissolved oxygen requirement[††] through aeration and agitation
	8. Anaerobic fermentation (obligate or facultative), depending on microorganisms	E.g. acetone butanol fermentation, ethanol fermentation, lactic acid fermentation
Nutrient metabolism	9. Conserved growth	Regulation of the substrate uptake
	10. Synchronous growth	Induction of synchronous cell division
	11. Diauxic growth	Biphasic pattern of growth
	12. Cryptic growth	Viable cells use lysed cells for growth (e.g. in mixed cultures)
	13. Solventogenic fermentation	E.g. acetone butanol fermentation, ethanol and glycerol fermentations
	14. Acidogenic fermentation	E.g. acetic acid, citric acid, lactic acid, propionic acid
	15. Homofermentation	Only one major end product
	16. Heterofermentation	More than one major end product
Number of inoculums	17. Mono fermentation	Only one culture is used for a specified purpose
	18. Dual fermentation	Multiple strains are used for the production of a particular metabolite (e.g. vinegar)
	19. Mixed fermentation	More than two organisms are involved such as in most natural fermentations

In **Table 1**, it is also imperative to note that the diverse factors determining the types of fermentation are didactically separated, but in reality, they are interdependent and not exclusive of one another. It is similarly important to note that different types of fermentation share the same examples of processes and products.

The term brewing is rigorously applied to the hot extraction of plant materials (e.g. coffee or tea brewing), although it is more usually applied to the entire process

---

[††]Oxygen is sparingly soluble in water and 6000 times less soluble than glucose at standard temperature and pressure [28].

of making beer (or lager) by the fermentation of an aqueous extract of (malted) barley (*Hordeum vulgare* L., Poaceae), containing essential oils and bitter resins of the dried female flowers (cones) of the hop (*Humulus lupulus* L., Cannabaceae) [6, 18, 27]. Lupulin (an early, generic, name) contains the hop α-acids and β-acids (**Figure 7**), essential oils and polyphenols that give the bitterness and flavor to the beer [30, 31]. Thus, beer is made of four main ingredients: malt (the germinated barley), yeasts (e.g. strains of *Saccharomyces carlsbergensis*), water and hops [32-34]. These ingredients allow brewers to create differentiated varieties of beer [32]. It remains true that the origin of the popular word beer is not well established, but it is closely related to the Latin word *bibere* (to drink) [35].

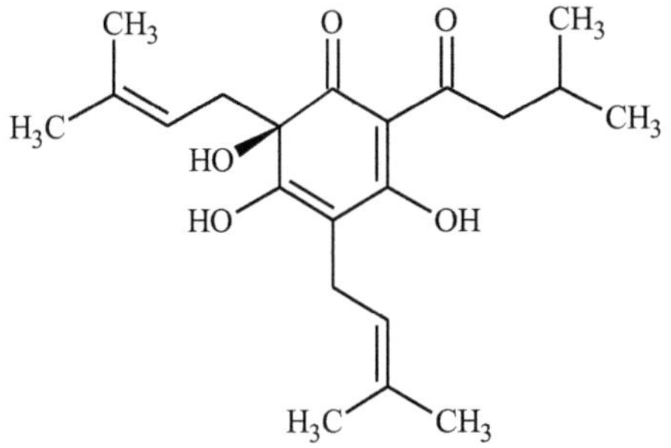

Figure 7:
Humulone: an example of α-acid responsible for the bitterness of beer. These acids have no bitter taste until they isomerize to form iso-α-acids [31].

2.2 Enzymes: the etymology, terminology and main characteristics

The etymon the closest to the current English term enzyme came from the German word *Enzym*, with its etymological root in the Late Greek, transliterated as *enzyme(os)*, meaning leavened [1, 36]. Most enzymes are protein polymers composed of the 20 naturally occurring amino acids plus the rarer selenocysteine[‡‡] and L-pyrrolysine (apart from ribozyme or RNA with catalytic activity) [27, 37-39].

[‡‡]Selenocysteine is referred to as "nature's 21st proteinogenic amino acid" and it can be readily incorporated into chemically synthesized peptides [38]. Its high reactivity is likely the reason why evolution has selected disulfide bridges over diselenide bridges to form covalent crosslinks in polypeptides [38].

Enzymes are catalysts, *i.e.* they are substances that without undergoing permanent alteration cause chemical reactions to proceed at faster rates [40, 41]. Histidine and cysteine are strongly catalysts in all classes of enzymes (EC 1, oxidoreductases; EC 2, transferases; EC 3, hydrolases; EC 4, lyases; EC 5, isomerases; EC 6, ligases) [37]. As the example of pectinesterase (EC 3.1.1.11) cited in page **17**, the first three levels (enzyme class, subclass and sub-subclass) broadly define the overall chemistry occurring and the serial number (the fourth level) generally defines the substrate specificity [37].

Enzymes catalyze an enormous array of chemical reactions, with regiospecificity and stereospecificity and under physiological conditions [37]. These vital functions are possible because enzymes are sophisticated and structured substances. Nevertheless, the enzyme structural complexity is such that the component parts may seem an ambiguous biochemical jargon (e.g. coenzyme, cofactor, prosthetic group, among others) [42]. These are fundamental concepts that deserved further consideration herein as follows.

Many regulatory enzymes have a quaternary structure, *i.e.* they are polymers composed of several (identical) monomer subunits, and they have a definite structural axis of symmetry, which is maintained during their interaction with substrate molecules (either inhibitors or activators) via the active (or catalytic) site [37, 43, 44]. For those enzymes that rely solely on amino acids for catalysis, the types of reactions are extremely narrow in scope and, for the most part, restricted to acid/base and electrophilic/nucleophilic reactions [41].

Thus, a holoenzyme is defined as the complete form of an enzyme, *i.e.* the complex of all protein subunits, the coenzyme and the prosthetic group [4, 27]. A prosthetic group, *i.e.* a covalently associated nonprotein constituent required for a particular function, may also be termed a coenzyme if it is directly involved in catalytic reaction (on the other hand, coenzyme and cofactor weakly bound to enzyme are, however, not classified as prosthetic groups) [42, 44]. At present, coenzyme is considered a synonym for cofactor (of organic or inorganic nature) by many authors [42, 44-47].

Additionally, when enzymes are in their cofactor-bound state, they are referred to as holoenzymes (active forms) while when they are in their unbound state, they are known as apoenzymes (inactive forms) – **Figure 8** [41]. Accordingly, some studies suggested that the holoenzyme, and not the apoenzyme isolated, could recognize selective substrates [48, 49]. Despite the promising emergence of synthetic biology[§§], the impact of cofactor manipulation on the activity of holoenzymes is still a poorly studied area [41, 50].

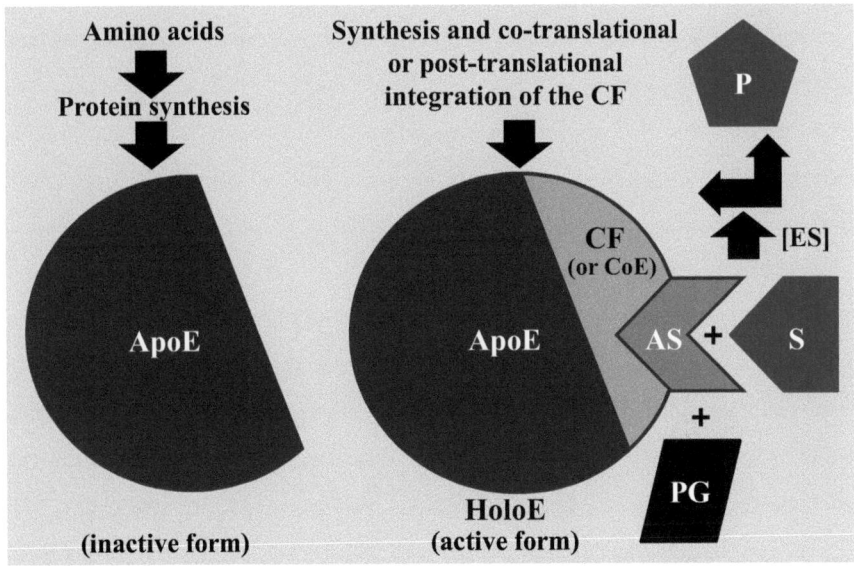

Figure 8:
Functional components of enzyme structure (adapted from [41]).
ApoE: apoenzyme; HoloE: holoenzyme; CF: cofactor; CoE: coenzyme; AS: active site; S: substrate; PG: prosthetic group; [ES]: enzyme-substrate complex; Double arrow: dissociation of the HoloE from the product (P).

It is worth noting that vitamins are usually functional components of enzyme systems common to all forms of life, from bacteria to human beings [51]. They

_____
[§§]Defined as the application of engineering principles to biology [50].

perfectly fit the definitions of coenzymes, cofactors and prosthetic groups such as biotin (or vitamin B_7), among others [41, 44, 52].

Moreover, isoforms are highly related gene products that perform essentially the same biological function [53]. However, they are different in amino acid sequence and electrophoretic mobilities, and they may be produced by different genes or by alternative splicing of RNA transcripts[***] from the same gene [27]. Isozymes or isoenzymes or yet isofunctional enzymes, are isoforms of an enzyme that act as their precursors, *i.e.,* in sum, they are different enzymes that catalyze the same reaction [27, 53, 54]. A more subtle aspect involves evolutionarily unrelated proteins that catalyze the same biochemical reactions, often referred to as analogous – as opposed to homologous – or more accurately non-homologous isofunctional enzymes such as cellulase (EC 3.2.1.4) and peroxidase (EC 1.11.1.7) [55]. Actually, sets of evolutionarily unrelated, non-homologous isofunctional enzymes were detected for a substantial fraction (up to 10%) of biochemical reactions and the adequate description of these enzymes is important for the practical tasks of metabolic reconstruction and enzyme classification [55].

Furthermore, enzymes are considered the perfect biotechnological tools for the formulation of foods, since they are specific to their substrates and selective for a particular type of molecular bonding [56-58]. Enzymes that diverge from this common statement are called promiscuous, *i.e.* enzymes with the ability to catalyze alternate chemical reactions besides the natural ones such as the peroxidases [58, 59]. Peroxidases are important enzymes in the food industry since they are related to food quality via browning processes and fiber formation in harvested fruits and vegetables [60]. This new research area has shown many potential applications from the laboratory to the industrial scale [58].

Enzymes used in the food industry are derived from culturable nonpathogenic microorganisms, edible plants and animal tissues; however, enzymes from

---

[***]The transcript that results from the alternative splicing does not contain all the exons compared to that of constitutive splicing, which generates a transcript containing all exons [27].

microorganisms are more useful than the other mentioned sources because they offer a wide variety of catalytic activities, high yields, easy genetic manipulation, regular supply and a rapid growth on cheap media [61]. For instance, *Bacillus subtilis* (e.g. strain MGB874) is attractive for industrial use for a variety of reasons (**Figure 3**), including its rapid growth rate, ability to secrete proteins into the medium [62]. These proteins are extracellular enzymes whose functions are external to the cell; they may be free or cell-bound [63]. It is not redundant to emphasize their commercial importance to the food industry.

2.3 The definition and classification of fermented foods

There are many definitions of fermented foods, but they all emphasize the palatable nature of these foodstuffs as well as their processing [64-67]. Therefore, fermented foods may be defined as food substrates invaded (or overgrown) by edible microorganisms whose enzymes, particularly amylases (EC 3.2.1), proteases (EC 3.4) and lipases (EC 3.1.1), hydrolyze polysaccharides, proteins and lipids to non-toxic products with flavors, aromas and textures attractive to human consumers [61, 64, 66, 68, 69].

On the other hand, the classification of fermented foods is based on more diversified criteria, e.g. the categories of foods (e.g. alcoholic beverages fermented by yeasts; vinegars fermented with *Acetobacter*), the classes of foods (e.g. cereal products; fish products), the resultant commodities (e.g. fermented starchy roots; fermented cereals), the pertinent functional basis (e.g. food for special occasions) [69]. It is noteworthy that fermented foods constitute the predominant category of ethnic foods, as further discussed in **CHAPTER 5** of this book [69].

REFERENCES

[1]   *Churchill's Medical Dictionary*, E. L. Becker, Lord Butterfield of Stechford, A. M. Harvey, R. H. Heptinstall, and L. Thomas, Eds., Churchill Livingstone, New York, NY, USA, 1989.

[2]   *Dorland's Illustrated Medical Dictionary*, D. A. Anderson, J. M. Patwell, K. Plaut, and K. McCullough, Eds., W. B. Saunders Company, Philadelphia, PA, USA, 1988.

[3]   J. Nielsen, "Fermentation monitoring, design and optimization," in *The Encyclopedia of Bioprocess Technology: Fermentation, Biocatalysis, and Bioseparation*, M. C. Flickinger and S. W. Drew, Eds., pp. 1147- 1156, John Wiley & Sons, New York, NY, USA, 1999.

[4]   K. R. Nill, *Glossary of Biotechnology Terms*, CRC Press, Boca Raton, FL, USA, 2002.

[5]   P. Robert, *Petit Robert 1*, Dictionnaires Le Robert, Paris, France, 1986.

[6]   N. N. Potter and J. H. Hotchkiss, *Food Science*, Aspen, MD, USA, 1998.

[7]   H. W. Atwater, "Bread and the principles of bread making," *Farmer's Bulletin*, no. 112, pp. 7-39, 1900.

[8]   C. Dow, *The Healing Power of Tea: Simple Teas & Tisanes to Remedy and Rejuvenate your Health*, Llewellyn Worldwide Ltd., Woodbury, MN, USA, 2014.

[9]   W. H. Stahl, "The chemistry of tea and tea manufacturing," in *Advances in Food Research*, C. O. Chichester, Ed., pp. 201-262, Academic Press, New York, NY, USA, 1962.

[10] G. Lian, A. Thiru, A. Parry, and Steve Moore, "CFD simulation of heat transfer and polyphenol oxidation during tea fermentation," *Computers and Electronics in Agriculture*, vol. 34, no. 1-3, 145-158, 2002.

[11] V. R. Sinija and H. N. Mishra, "Green tea: health benefits," *Journal of Nutritional & Environmental Medicine*, vol. 17, no. 4, pp. 232-242, 2008.

[12] S. R. Evangelista, M. G. da C. P. Miguel, C. de S. Cordeiro, C. F. Silva, A. C. M. Pinheiro, and R. F. Schwan, "Inoculation of starter cultures in a semi-dry coffee (*Coffea arabica*) fermentation process," *Food Microbiology*, vol. 44, pp. 87-95, 2014.

[13] W. Masoud, L. B. Cesar, L. Jespersen, and M. Jakobsen, "Yeast involved in fermentation of *Coffea arabica* in East Africa determined by genotyping and by direct denaturing gradient gel electrophoresis," *Yeast*, vol. 21, no. 7, pp. 549-556, 2004.

[14] R. V. M. dos Santos, H. D. Vieira, F. M. Borém, and E. P. Isquierdo, "A decision support system to aid the calculation of the cost of the post-harvest processing of coffee," *Coffee Science*, vol. 8, no. 4, pp. 439-449, 2013.

[15] C. Stannard, "Development and use of microbiological criteria for foods," *Food Science and Technology Today*, vol. 11, no. 3, pp. 137-176, 1997.

[16] P. Fellows, *Food Processing Technology: Principles and Practice*, CRC Press, Boca Raton, FL, USA, 2000.

[17] J. Diao and M. S. Hasson, "Crystal structure of butyrate kinase 2 from *Thermotoga maritima*, a member of the ASKHA superfamily of phosphotransferases," *Journal of Bacteriology*, vol. 191, no. 8, pp. 2521-2529, 2009.

[18] P. Singleton and D. Sainsbury, *Dictionary of Microbiology and Molecular Biology*, John Wiley & Sons, London, UK, 2006.

[19] M. K. Jain and J. G. Zeikus, "Anaerobes, industrial uses," in *The Encyclopedia of Bioprocess Technology: Fermentation, Biocatalysis, and Bioseparation*, M. C. Flickinger and S. W. Drew, Eds., pp. 150-170, John Wiley & Sons, New York, NY, USA, 1999.

[20] S. Varadarajan and D. J. Miller, "Catalytic upgrading of fermentation-derived organic acids," *Biotechnology Progress*, vol. 15, no. 5, pp. 845-854, 1999.

[21] M. de Barrosa, S. Freitas, G. S. Padilha, and R. M. Alegre, "Biotechnological production of succinic acid by *Actinobacillus succinogenes* using different substrate," *Chemical Engineering Transactions*, vol. 32, pp. 985-990, 2013.

[22]   H. Song and S. Y. Lee, "Production of succinic acid by bacterial fermentation," *Enzyme and Microbial Technology*, vol. 39, no. 3, pp. 352-361, 2006.

[23]   W. Gratzer, *Giant Molecules: From Nylon to Nanotubes*, Oxford University Press Inc., New York, NY, USA, 2009.

[24]   T. Ghaffar, M. Irshad, Z. Anwar, T. Aqil, Z. Zulifqar, A. Tariq, M. Kamran, N. Ehsan, and S. Mehmood, "Recent trends in lactic acid biotechnology: a brief review on production to purification," *Journal of Radiation Research and Applied Sciences*, vol. 7, no. 2, pp. 222-229, 2014.

[25]   D. M. Vasudevan, S. Sreekumari, and K. Vaidyanathan, *Textbook of Biochemistry for Dental Students*, Jaypee Brothers Medical Publishers, New Delhi, India, 2011.

[26]   M. N. Haque, R. Chowdhury, K. M. S. Islam, and M. A. Akbar, "Propionic acid is an alternative to antibiotics in poultry diet," *Bangladesh Journal of Animal Science*, vol. 38, no. 1-2, pp. 115-122, 2009.

[27]   P.-S. Juo, *Concise Dictionary of Biomedicine and Molecular Biology*, CRC Press, Boca Raton, FL, USA, 2002.

[28]   K. Jayachandran, I. C. Nair, T. S. Swapna, and A. Sabu, "Fermentation of food processing by-products," in *Valorization of Food Processing By-Products*, M. Chandrasekaran, Ed., pp. 203-232, CRC Press, Boca Raton, FL, USA, 2012.

[29]   C. Lüdecke, K. D. Jandt, D. Siegismund, M. J. Kujau, E. Zang, M. Rettenmayr, J. Bossert, and M. Roth, "Reproducible biofilm cultivation of chemostat-grown *Escherichia coli* and investigation of bacterial adhesion on biomaterials using a non-constant-depth film fermenter," *PLoS ONE*, vol. 9, no. 1: e84837, 2014.

[30]   A. M. Fortes, F. Santos, and M. S. Pais, "Organogenic nodule formation in hop: a tool to study morphogenesis in plants with biotechnological and medicinal applications," *Journal of Biomedicine and Biotechnology*, vol. 2010, 2010, 16 p.

[31]   J. E. Page and J. Nagel, "Biosynthesis of terpenophenolic metabolites in hop and *Cannabis*," *Recent Advances in Phytochemistry*, vol. 40, pp. 179-210, 2006.

[32]   G. Gabrielyan, J. J. McCluskey, T. L. Marsh, and C. F. Ross, "Willingness to pay for sensory attributes in beer," *Agricultural and Resource Economics Review*, vol. 43, no. 1, pp 125-139, 2014.

[33]   S. Yoshida, K. Hashimoto, E. Shimada, T. Ishiguro, T. Minato, S. Mizutani, H. Yoshimoto, K. Tashiro, S. Kuhara, and O. Kobayashi, "Identification of bottom-fermenting yeast genes expressed during lager beer fermentation," *Yeast*, vol. 24, pp. 599-606, 2007.

[34]   M. Henderson, "Whiskey," in *Handbook of Food and Beverage Fermentation Technology*, Y. H. Hui, L. Meunier-Goddik, A. S. Hansen, J. Josephsen, W.-K. Nip, P. S. Stanfield, and Fidel Toldá, Eds., pp. 879-890, Marcel Dekker, Inc., New York, NY, USA, 2004.

[35]   D. De Keukeleire, J. Vindevogel, R. Szücs, and P. Sandra, "The history and analytical chemistry of beer bitter acids," *Trends in Analytical Chemistry*, vol. 77, no. 8, pp. 275-280, 1992.

[36]   R. A. Copeland, *Enzymes: A Practical Introduction to Structure, Mechanism, and Data Analysis*, Wiley-VCH, New York, NY, USA, 2000.

[37]   G. L. Holliday, J. D. Fischer, J. B. O. Mitchell, and J. M. Thornton, "Characterizing the complexity of enzymes on the basis of their mechanisms and structures with a bio-computational analysis," *FEBS Journal*, vol. 278, no. 20, pp. 3835-3845, 2011.

[38]   A. M. Steiner, K. J. Woycechowsky, B. M. Olivera, and G. Bulaj, "Reagentless oxidative folding of disulfide-rich peptides is catalyzed by an intramolecular diselenide," *Angewandte Chemie International Edition in English*, vol. 51, no. 23, pp. 5580-5584, 2012.

[39]   C. H. Lee, J. H. Kim, and S.-W. Lee, "Prospects for nucleic acid-based therapeutics against hepatitis C virus," *World Journal of Gastroenterology*, vol. 19, no. 47, pp. 8949-8962, 2013.

[40]   M. A. Tabatabai, "Soil enzymes," in *Methods of Soil Analysis: Part 2—*

Microbiological and Biochemical Properties, P. S. Bottomley, J .S. Angle, and R. W. Weaver, Eds., pp. 775-833, The Soil Science Society of America, Inc., Madison, WI, USA, 1994.

[41] M. K. Akhtar and P. R. Jones, "Cofactor engineering for enhancing the flux of metabolic pathways," *Frontiers in Bioengineering and Biotechnology*, vol. 2, no. 30, pp. 1-6, 2014.

[42] O. H. Hashim and N. A. Adnan, "Coenzyme, cofactor and prosthetic group – ambiguous biochemical jargon," *Biochemical Education*, vol. 22, no. 2, pp. 93-94, 1994.

[43] R. E. Bulger and J. M. Strum, "Energy production and use," in *The Functioning Cytoplasm*," R. E. Bulger and J. M. Strum, Eds., pp. 17-31, Springer, New York, NY, USA, 1974.

[44] D. L. Nelson and M. M. Cox, *Lehninger Principles of Biochemistry*, W. H. Freeman and Company, New York, NY, USA, 2013.

[45] J. D. Mauseth, *An Introduction to Plant Biology,* Jones & Barlett Learning, Burlington, MA, USA, 2014.

[46] A. D. Lawrence, S. L. Taylor, A. Scott, M. L. Rowe, C. M. Johnson, S. E. J. Rigby, M. A. Geeves, R. W. Pickersgill, M. J. Howard, and M. J. Warren, "FAD binding, cobinamide binding and active site communication in the corrin reductase (CobR)," *Bioscience Reports*, vol. 34, no. 4, pp. 345-355, 2014.

[47] J. Song, Z. Hong, R. K. Nagarale, and W. Shin, "Simple preparation of diaphorase/polysiloxane viologen polymer-modified electrode for sensing NAD and NADH," *Journal of Electrochemical Science and Technology*, vol. 2, no. 3, pp. 163-167, 2011.

[48] S. Angelaccio, R. Florio, V. Consalvi, G. Festa, and S. Pascarella, "Serine hydroxymethyltransferase from the cold adapted microorganism *Psychromonas ingrahamii*: a low temperature active enzyme with broad substrate specificity," *International Journal of Molecular Sciences*, vol. 13, no. 2, pp. 1314-1326, 2012.

46

[49] D. Biswas, V. Pandya, A. K. Singh, A. K. Mondal, and S. Kumaran, "Co-factor binding confers substrate specificity to xylose reductase from *Debaryomyces hansenii*," *PLoS ONE*, vol. 7, no. 9: e45525, 2012.

[50] L. Zhu, Y. Zhu, Y. Zhang, and Y. Li, "Engineering the robustness of industrial microbes through synthetic biology," *Trends in Microbiology*, vol. 20, no. 2, pp. 94-101, 2012.

[51] V. P. Sydenstricker, "The impact of vitamin research upon medical practice," *Proceedings of the Nutrition Society*, vol. 12, no. 03, pp. 256-269, 1953.

[52] S. R. Manning and J. W. La Claire II, "Prymnesins: toxic metabolites of the golden alga, *Prymnesium parvum* Carter (Haptophyta)," *Marine Drugs*, vol. 8, no. 3, pp. 678-704, 2010.

[53] P. W. Gunning, "Protein isoforms and isozymes," in *Encyclopedia of Life Sciences*, pp.1-5, John Wiley & Sons, Ltd., 2005.

[54] R. E. Viola, "The central enzymes of the aspartate family of amino acid biosynthesis," *Accounts of Chemical Research*, vol. 34, no. 5, pp. 339-349, 2001.

[55] M. V. Omelchenko, M. Y. Galperin, Y. I. Wolf, and E. V. Koonin, "Non-homologous isofunctional enzymes: a systematic analysis of alternative solutions in enzyme evolution," *Biology Direct*, vol. 5, no. 31, 2010.

[56] S. Couri, Y. Park, G. Pastore, and A. Domingos, "Enzimas na produção de alimentos e bebidas," in *Enzimas em Biotecnologia: Produção, Aplicações e Mercado*, E. P. S. Bom, M. A. Ferrara, and M. L. Corvo, Eds., pp. 153-177, Interciência Ltda., Rio de Janeiro, RJ, Brazil, 2008.

[57] M. A. Malajovich, *Biotecnologia*, Axcel Books do Brasil, Rio de Janeiro, RJ, Brazil, 2004.

[58] K. Hult and P. Berglund, "Enzyme promiscuity: mechanism and applications," *Trends in Biotechnology*, vol. 25, no. 5, pp. 231-238, 2007.

[59] F. G. Mutti, M. Lara, M. Kroutil, and W. Kroutil, "Ostensible enzyme promiscuity: alkene cleavage by peroxidases," *Chemistry*, vol. 16, no. 47, pp. 14142-

1448, 2010.

[60] V. Guida, M. Cantarella, A. Chambery, M. C. Mezzacapo, A. Parente, N. Landi, V. Severino, and A. Di Maro, "Purification and characterization of novel cationic peroxidases from *Asparagus acutifolius* L. with biotechnological applications," *Molecular Biotechnology*, vol. 56, no. 8, pp. 738-746, 2014.

[61] Y.-H. P. Hsieh and J. A. Ofori, "Innovations in food technology for health," *Asia Pacific Journal of Clinical Nutrition*, vol. 16, supplement 1, pp. 65-73, 2007.

[62] K. Manabe, Y. Kageyama, T. Morimoto, E. Shimizu, H. Takahashi, S. Kanaya, K. Ara, K. Ozaki, and N. Ogasawara, "Improved production of secreted heterologous enzyme in *Bacillus subtilis* strain MGB874 via modification of glutamate metabolism and growth conditions," *Microbial Cell Factories*, vol. 12, no. 18, 2013.

[63] J. Marxsen, "Bacteria and Fungi," in *Central European Stream Ecosystems: The Long Term Study of the Breitenbach*, R. Wagner, J. Marxsen, P. Zwick, and E. J. Cox, Eds., pp. 131-194, Wiley-VCH Verlag GmbH & Co. KGaA, Weinheim, Germany, 2011.

[64] A. T. Adesulu and K. O. Awojobi, "Enhancing sustainable development through indigenous fermented food products in Nigeria," *African Journal of Microbiology Research*, vol. 8, no. 12, pp. 1338-1343, 2014.

[65] A. A. O. Ogunshe and K. O. Olasugba, "Microbial loads and incidence of food-borne indicator bacteria in most popular indigenous fermented food condiments from middle-belt and southwestern Nigeria," *African Journal of Microbiology Research*, vol. 2, no. 12, pp. 332-339, 2008.

[66] Z. Kohajdová, J. Karovičová, and M. Greifová, "Analytical and organoleptic profiles of lactic acid-fermented cucumber juice with addition of onion juice," *Journal of Food and Nutrition Research*, vol. 46, no. 3, pp. 105-111, 2007.

[67] P. M. Kenneally, R. G. Leuschner, and E. K. Arendt, "Evaluation of the lipolytic activity of starter cultures for meat fermentation purposes," *Journal of Applied Microbiology*, vol. 84, no. 5, pp. 839-846, 1998.

[68]   J. Karovičová and Z. Kohajdová, "Lactic acid fermented vegetable juices," *Horticultural Science* (Prague), vol. 30, no. 4, pp. 152-158, 2003.

[69]   K. H. Steinkraus, "Classification of fermented foods: worldwide review of household fermentation techniques," *Food Control*, vol. 8, no. 5/6, pp. 311-317, 1997.

CHAPTER 3

THE MICROBIOLOGICAL AND
BIOCHEMICAL (ENZYMOLOGICAL)
BASES OF THE MAJOR FERMENTATION PATHWAYS

Carbohydrates, such as glucose (**Figure 9**), are one of the most important sources of energy (adenosine triphosphate or ATP) for most living organisms [1, 2]. It is in part because the enzymes involved in the carbohydrate breakdown (glycolysis) are among the most highly conserved proteins in evolution [3-9]. Actually, glycolytic enzymes of vertebrates are similar (amino acid sequence, three-dimensional structure) to their homologs in yeasts and glycolysis differs among species only in the details of its regulation, and in the subsequent metabolic fate of the pyruvic acid generated [1]. Glycolysis[†††] is commonly described as an anaerobic fermentation that, in turn, is an expression used to refer to pathways by which organisms extract ATP from high-energy fuels in the absence of molecular oxygen (O_2) [10-16]. Despite yielding only a fraction of the energy available from the complete combustion of glucose, this non-oxygen requiring metabolic pathway represents an evolutionary advantage for many organisms (e.g. plants), especially obligate anaerobes (e.g. *Clostridium perfringens, Spirochaeta litoralis*), and explains in some measure the extensive use of anaerobic fermentation by unicellular organisms and even by multicellular organisms [17-19]. For instance, this type of fermentation is one of the major metabolic adaptations that certain crops (e.g. rice) assume when they are submerged or faced with lack of oxygen [19]. Nonetheless, most pertinent to this

---

[†††]Under physiological conditions, glycolysis occurs in the cytoplasm outside of the mitochondrion [12]. Glycolysis yields two ATP molecules per glucose molecule whereas oxidative phosphorylation produces up to 36 ATP molecules per glucose molecule [1, 13]. Glycosomes (similar to mammal peroxisomes) constitute the sites of glycolysis in certain protozoa (e.g. genus *Leishmania*) [14]. The Warburg effect is the observation that the metabolism of most tumor cells is characterized by increased glycolysis that is maintained even during conditions of high oxygen tension (*i.e.* aerobic glycolysis), followed by elevated lactate production levels [15]. Thus, except this effect, glycolysis is anaerobic (redundant phrase) [16].

book are the fermentation processes discussed further ahead.

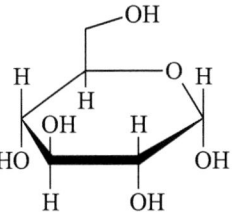

Figure 9:
Glucose (α-D-glucopyranose).
This closed structure is less prone to glycate proteins [18] – see subsection **(4.2.5)**.

3.1 Alcoholic fermentation

Alcoholic fermentation is a polymicrobial process (bacteria, yeasts, molds) by which carbohydrates are converted into two molecules of ethanol (**Figure 10**) and two of carbon dioxide under anaerobic conditions [17, 20-30]. More precisely, pyruvate decarboxylase (EC 4.1.1.1) is a key enzyme in alcoholic fermentation in yeasts and converts pyruvate into acetaldehyde (or ethanal) and carbon dioxide *en route* to ethanol [23, 31-35]. This decarboxylation depends on the availability of essential cofactors such as thiamine pyrophosphate and magnesium [33, 34, 36, 37]. In opposition, if the fermentation products are not preserved under anaerobic conditions, bacteria belonging to the genus *Acetobacter* may oxidize portions of the ethanol to acetic acid [38]. Thus, the end products of the fermentation processes may be the substrates of others, depending e.g. on the growth conditions of the microorganisms [18, 39, 40]. **Figure 11**.

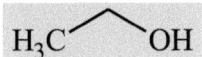

Figure 10:
Ethanol: its content is distinctively variable in fermented beverages.
However, a standard drink of beer was defined as 330 mL, containing 4.6% (vol/vol) ethanol corresponding to 12 g/drink, according to Danish standards [29]. The "zero tolerance policy" lowered the legally acceptable blood alcohol concentration for drivers under the age of 21 from 0.08 or 0.10 g/dL down to zero in the United States [30].

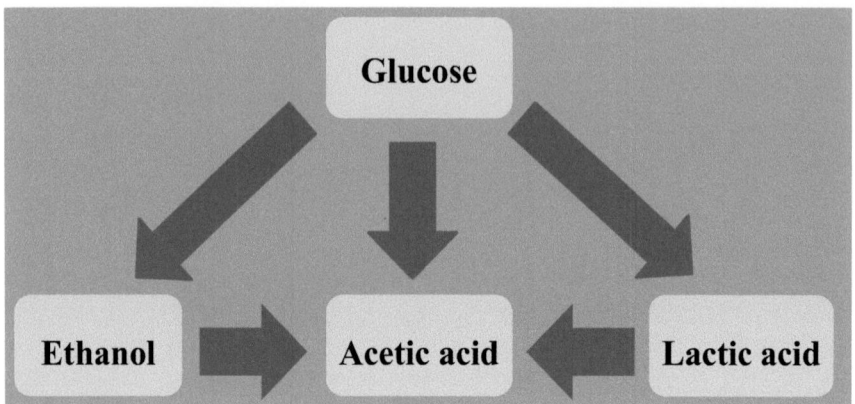

Figure 11:
Glucose: a rich source of by-products during fermentation (adapted from [18, 39-43]).
The glucose metabolic fate during fermentation depends on many of the factors discussed in **Table 1**. Microorganisms are obviously pivotal in the different processes. Acetogens convert glucose to acetic acid [41]. When there is abundance in glucose, or microorganisms (or certain species) are dominant (e.g. *Lactobacillus* or *L. buchneri*), glucose is converted to lactic acid, or to acetic acid [17, 40, 42, 43].

The previously mentioned baker's yeast *S. cerevisiae* (**CHAPTER 1**) is a microorganism with a remarkable metabolic flexibility, being able to grow fermentatively under strict anaerobic conditions as well as to use an entirely respiratory metabolism when sufficient oxygen is available [44]. In fact, *S. cerevisiae* has been shown to be metabolically versatile in other ways. Besides pyruvate decarboxylase, it possesses genes encoding alcohol dehydrogenase (EC 1.1.1.1) key isozymes (e.g. *ADH1*), which catalyze the reduction of acetaldehyde (or ethanal) to ethanol, oxidize NADH and provide NAD+ in the last step of the ethanol production pathway (e.g. the decrease in ethanol production by *ADH1* deletion alone can be partially compensated by the upregulation of other isozyme genes) [45, 46]. *S. cerevisiae* utilizes external thiamine for the production of thiamine diphosphate, but can also synthesize thiamine *de novo* (*i.e.* from the beginning) from hydroxyethylthiazole and hydroxy-methylpyrimidine [47-49].

Cellular compartmentalization is a key phenomenon to carry out orderly and controlled reactions [50]. In prokaryotic cells (e.g. bacteria), the chromosomes are generally 1000 times longer than the cells in which they reside and concurrent replication, segregation and transcription/translation of this crowded mass of DNA pose a challenging organizational problem [51]. On the other hand, eukaryotic cells (e.g. yeasts) compartmentalize biochemical processes in different organelles, often relying on metabolic cycles to shuttle reducing equivalents across intracellular membranes, which is critical for numerous functions [52].

In *S. cerevisiae*, the nucleus contains 16 relatively small chromosomes, comprising between 230 and 1,500 kb of DNA, plus 100-200 copies of ribosomal genes (rDNA) encompassing 1-2 Mb [53, 54]. *S. cerevisiae* increases in number by budding and some structural components are the cell wall, periplasm, plasma membrane, invaginations, bud scar, cytosol, mitochondrion, endoplasmic reticulum, Golgi apparatus, secretory vesicles, vacuoles and peroxisomes (**Figure 12**) [55]. In *S. cerevisiae*, proteins encoding the isozymes *ADH1*, *ADH2* and *ADH5* are localized to the cytoplasm, and those encoding *ADH3* and *ADH4* are localized to the mitochondria [56].

Additional molecular biology studies have partially clarified the underlying mechanisms of glucose transport in many eukaryotic species (humans, plants), including *S. cerevisiae* [57, 58]. It is worth noting that yeast cells show a remarkable preference for glucose (and closely related sugars) as a carbon source because these sugars can be converted rapidly to ethanol and the accumulation of ethanol in the medium gives a selective advantage to the highly ethanol-tolerant yeast cells [57]. This significant production of ethanol could be a waste product toxic to other microorganisms and could constitute an evolutionary mechanism favoring output over efficiency in the competition for nutrients [59].

S. cerevisiae is the most well established fermentation yeast for large scale alcoholic fermentation of the hexose (six-carbon) sugars (glucose, mannose and galactose); although it does not naturally metabolize the pentose (five-carbon) sugars (xylose and arabinose) without biotechnological intervention, being classified as a non-xylose-utilizing organism [60-63]. It is pertinent to note that the cryotolerant strains of the yeast *Saccharomyces bayanus* – which is distantly related to the cultured yeast *S. cerevisiae* – were found in winemaking and became the second most important yeast for basic and applied studies [64].

In addition, ale beers are made using selected strains of *S. cerevisiae*, otherwise known as the "ale" or "top-fermenting" yeasts [65]. One of the distinctive technological traits of ale yeasts is their ability to produce high concentrations of esters, compounds that impart a fruity aroma to beer [66]. Lager beers are fermented by *Saccharomyces pastorianus* (formerly called *S. carlsbergensis*), also known as the "lager" or "bottom-fermenting" yeasts [65]. Among the distinctive physiological characteristics of lager brewing strains are the ability to use melibiose (a disaccharide containing galactose and glucose)[‡‡‡] as a carbon source, especially the Malaysian strains, the ability to ferment well at low temperatures (8-10°C), the inability to grow above 37°C, the presence of an active system in the cell for fructose transport [66-68].

Alcoholic bacteria (e.g. *Zymomonas mobilis*) also have the potential to be used for alcoholic fermentation at the industrial scale, since they exhibit promising abilities to transform sugars into alcohol and carbon dioxide, at conditions similar to those required by yeasts [69].

To close, it is important to remark that molds are considered mostly for their spoilage role, especially contamination of the vats where wine or such beverages are stored for maturation [70]. The lactic acid bacteria, although more responsible for spoilage rather than any beneficial effect on the wine, may play a significant role in the malolactic fermentation by lowering the excessive acidity of wines, bringing biological stability and improving the organoleptic characteristics of the product [70, 71].

Alcoholic fermentation is known in detail as outlined in **Figure 12**. *S.*

[‡‡‡]The loss of melibiose utilization associates with complete loss of α-galactosidase (EC 3.2.1.22) or melibiase, catalyzing the conversion of that disaccharide in the two hexoses abovementioned [68].

cerevisiae has an ellipsoidal shape, but the emphasis was given to its compartmentalization [55].

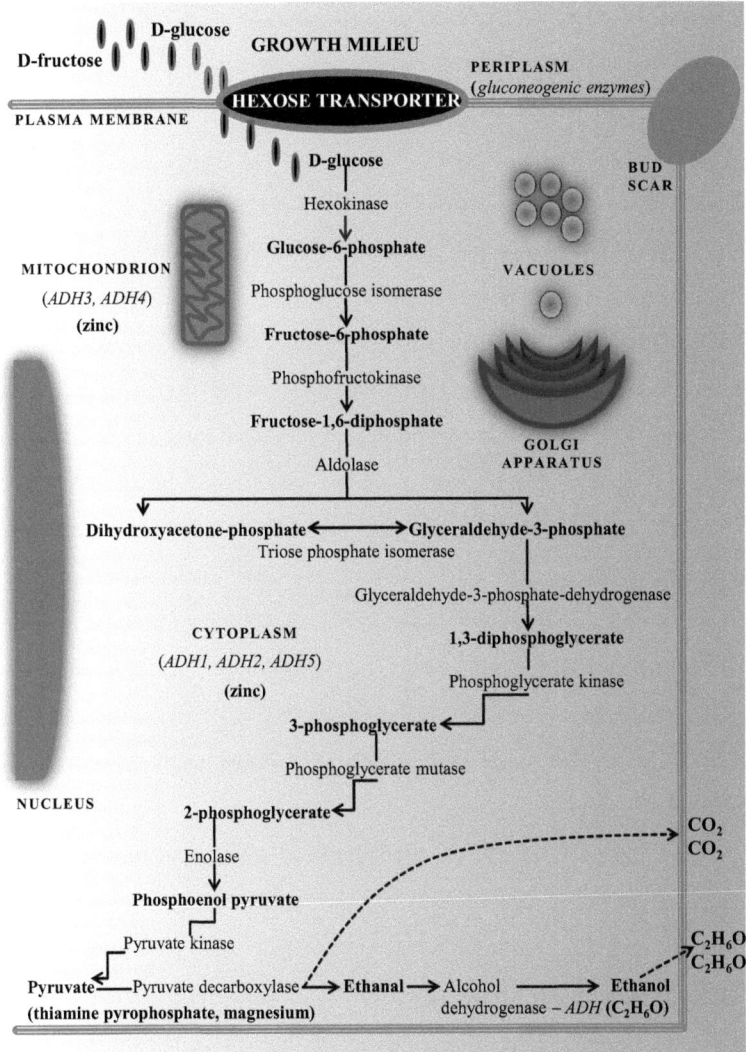

Figure 12:
Glycolysis and alcoholic fermentation in *S. cerevisiae* (adapted from [1, 18, 23, 33, 34, 36, 37, 55, 72-74]). Yeasts are facultative anaerobic microorganisms because they possess the genetic equipment for metabolizing sugars in two pathways, aerobically (via mitochondrial respiration) or anaerobically (via cytoplasmic fermentation) [23]. Yeasts favor glyoxylate cycle with acetyl-CoA to form malate (bypassing the normal Krebs cycle sequence) and circumvent the "irreversible glycolytic steps" via gluconeogenesis (glucose-6-phosphate synthesis via noncarbohydrate precursors) to affect their aging negatively [67, 72, 73].

Some aspects of **Figure 12** should be further discussed. Pyruvate is a three-carbone molecule that is, however, located at a major junction of assimilatory and dissimilatory reactions as well as at the branch-point between respiratory dissimilation of sugars and alcoholic fermentation in yeasts [33, 75, 76]. Pyruvate transport into the mitochondrial matrix is necessary prior to its decarboxylation into acetyl-CoA (via pyruvate dehydrogenase complex), which feeds the reducing equivalent generating the Krebs cycle or citric acid cycle or yet tricarboxylic acid cycle [1, 18, 33, 76]. As the initial reduction of pyruvate to ethanal is a nonreversible step, ethanal, rather than being converted back to pyruvate, is converted to acetate by aldehyde dehydrogenase (primarily in the mitochondria) before final oxidization into carbon dioxide and water [33, 77]. Transgenic yeasts expressing exogenous L-lactate dehydrogenase (EC 1.1.1.27) could produce lactic acid from pyruvic acid, but a considerable amount of ethanol was produced concurrently because *S. cerevisiae* predominantly produces ethanol under anaerobic conditions [33, 78]. It was also demonstrated that the (active or inactive) mitochondrial status of the yeast affects organic acid production (e.g. succinic and malic acids) during alcoholic fermentation [79]. This data is clearly important both in theory and in practice. Organic acids play a major role (taste, flavor) of alcoholic beverages (e.g. sake, the traditional Japanese rice wine) and most of those molecules are produced by *S. cerevisiae* during alcoholic fermentation [79].

It is also necessary to make a brief review of some regulatory phenomena of fermentative sugar metabolism in yeasts. The Pasteur effect[§§§] is defined as the suppression of alcoholic fermentation in the presence of oxygen and involves the inactivation of 6-phosphofructokinase by oxidized cytochrome [17, 33, 49, 67, 80]. The short-term Crabtree effect is defined as the immediate occurrence of aerobic alcoholic fermentation in response to provision of a pulse of excess sugar to sugar-limited yeast cultures [33, 81, 82]. The long-term Crabtree effect is defined as the

---

[§§§]Named after Herbert Grace Crabtree, English physician and biochemist, who reported it in 1929 [49].

aerobic alcoholic fermentation at high growth rates, irrespective of the mode of cultivation (growth under sugar limitation or growth with excess sugar) [33]. The Kluyver effect is defined as the absence of alcoholic fermentation during oxygen-limited growth on a sugar (often a disaccharide), even though glucose is readily fermented [33]. It is probably caused mainly by slower uptake of sugar anaerobically [80]. In the Custers effect**** (or negative Pasteur effect), yeasts of the genus *Brettanomyces* (or its teleomorph form *Dekkera*) ferment D-glucose to ethanol and CO_2 faster in aerobic than in anaerobic conditions [17, 33, 80, 83, 84]. The contamination of wines by these spoilage yeasts is an actual and very important problem in many wineries worldwide [84]. It is worth noting that the Kluyver effect does not occur in *S. cerevisiae* and the Pasteur effect is insignificant in this yeast [33, 80].

As previously mentioned and observed in other organisms, the metabolic versatility of *S. cerevisiae* is due to its powerful enzymatic equipment, including the compensatory activity of its isozymes†††† [85-87]. It performs glycolysis (the anaerobic breakdown of glucose to pyruvate), the Krebs cycle (a series of enzymatic reactions in aerobic organisms that involves oxidative metabolism of acetyl units and produces high-energy phosphate compounds, which serve as the main source of cellular energy), gluconeogenesis (the formation of glucose from compounds that are not carbohydrates, e.g. amino acids, glycerol or lactate) and the synthesis of glycogen (a major intracellular reserve of glucose as a polymer) [67, 80, 88, 89].

Lastly, substrate cycles, also known as futile cycles, are cyclic metabolic routes that dissipate energy by hydrolyzing cofactors such as ATP [1, 90]. The presence of many futile cycles (more in humans than in yeasts) can be explained by the existence of several enzymes catalyzing similar reactions with opposite directions [91]. For instance, without regulation of the enzyme activity of the two reciprocal pathways

---

****This phenomenon was first described by Custers, who named it "negative Pasteur effect" (1940) and Scheffers introduced the term "Custers effect" (1966) [83].
††††Alcohol dehydrogenase key isozymes are zinc-dependent, as outlined in **Figure 12** [86, 87].

(glycolysis and gluconeogenesis), an energy-wasting (futile cycle) would ensue, between phosphofructokinase-1 (EC 2.7.1.11) and fructose-bisphosphatase (EC 3.1.3.11) [80]. Hence, futile cycles may be prevented by mechanisms that inactivate either the forward or the reverse reactions, generally by a covalent modification of the bifunctional enzyme, such as phosphorylation, or by binding of a regulatory protein [92]. Otherwise, they could be used to increase ethanol yield in microorganisms, but the studies in this area are still scarce [93, 94].

3.2 Acetic acid fermentation

Acetic acid produced by primary microbial metabolism is denominated as acetic acid fermentation [95]. Accordingly, acetic acid fermentation is the biochemical process by which acetic acid is produced from ethanol by acetic acid bacteria under strict aerobic conditions [95-97]. Alcohol dehydrogenase and aldehyde dehydrogenase are key membrane-bound enzymes in the acetic acid fermentation [96-99]. The cloning and the sequence analyses of the genes that codify these two enzymes of *Acetobacter aceti* (72 and 50 kDa, respectively) permitted to understand their oxidative activities (additionally, the genes responsible for resistance to acetic acid **(Figure 13)** and ethanol of mutant acetic bacteria have also been studied) [96].

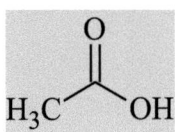

Figure 13:
Acetic acid: the major organic acid with concentrations up to 10 wt%, depending on the feedstock and process conditions applied in the pyrolysis process [100]. As already mentioned, acetic acid is also a by-product of *S. cerevisiae* alcoholic fermentation and together with high concentrations of ethanol and other toxic metabolites, acetic acid may contribute to fermentation arrest and reduced ethanol productivity [101, 102].

Acetic acid bacteria are related to the class Alphaproteobacteria, the family Acetobacteraceae that currently includes twelve genera, *i.e. Acetobacter, Gluconobacter, Acidomonas, Gluconacetobacter, Asaia, Kozakia, Swaminathania, Saccharibacter, Neoasaia, Granulibacter, Tanticharoenia* and *Ameyamaea* [103, 104]. These ubiquitous microorganisms are well adapted to sugar and ethanol rich environments [105]. Although they are predominantly found in fruits and flowers, being responsible for the decay of these natural resources [103]. The genus *Acetobacter* is the most significant for the industrial production of vinegar [103]. It is composed of aerobic, Gram-negative, non-sporeforming, acidophilic rod-shaped bacteria, which utilize a wide variety of organic compounds [106]. Interestingly, the alcohol dehydrogenase activity of *Acetobacter* is more stable under acetic conditions than that of *Gluconobacter*, explaining the higher production of acetic acid by *Acetobacter* [107]. As a counterpart, the ability to convert intermediate metabolites (e.g. D-xylulose) to xylitol[‡‡‡‡] is only marginally greater in strains of *Acetobacter* than in those of *Gluconobacter* [108, 109].

Finally, it is important to note that acetic acid may act as bacteriostatic or bactericidal (in dependence of its concentration) and even be a stronger preservative than ethanol, but both the alcoholic and the acetic acid fermentations yield foods that are generally safe [38]. Besides, one study in animals (rabbits) demonstrated that vinegar (5 and 10 mL) with hypercholesterolemic diet for 8 weeks caused significant reduction of several risk factors for atherosclerosis progression (e.g. C-reactive protein, fibrinogen, factor VII, apolipoprotein B (ApoB$_{100}$), total cholesterol), but future studies should focus on determining similar effects of vinegar on humans [110].

[‡‡‡‡]Xylitol is one of the alternative natural sweeteners, belonging to a group of sugar alcohols [109].

59

3.3 Lactic acid fermentation

Certainly, lactic acid fermentation is the simplest and safest way of preserving food and has probably always been used by humans [111]. Lactic acid fermentation is widely used in tropical countries for various reasons: methodological simplicity, low-cost and enhancement not only of food quality, but also of food shelf life and safety by its metabolic products, e.g. lactic acid (Figure 14), bacteriocins, among others [112-116]. Lactic acid fermentation is a one-step conversion from pyruvate, which is catalyzed by L-lactate dehydrogenase (EC 1.1.1.27) with a concomitant oxidation of nicotinamide adenine dinucleotide [34].

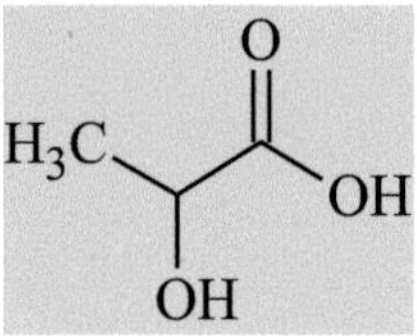

Figure 14:
Lactic acid (2-hydroxypropanoic acid).
Also known as milk acid, it is a chemical compound that plays a role in several biochemical processes and is widely distributed in nature [114, 115]. In 1881, Fermi obtained lactic acid by fermentation, resulting in its industrial production [115]. Microorganisms that synthesize lactic acid include lactic acid producing bacteria (e.g. *Lactobacillus* sp.) and filamentous fungi (e.g. *Rhizopus* sp.), which can convert 1 g glucose into 1 g lactic acid by homolactic fermentation pathway or 0.5 g lactic acid, 0.26 g ethanol and 0.24 g CO_2 by heterolactic fermentation pathway [116].

Thus, lactic acid bacteria normally carry out lactic acid fermentation and comprise various species of the genera *Carnobacterium, Enterococcus, Lactobacillus, Lactococcus, Leuconostoc, Oenococcus, Pediococcus, Streptococcus, Tetragenococcus, Vagococcus* and *Weissella* [117, 118]. These are anaerobic Gram-positive bacteria widely distributed in the nature, generally regarded as safe, extensively used in agriculture and in the food industry to produce fermented foods from plants and animals [118-124]. They are characterized as non-sporeforming, usually non-motile cocci, coccobacilli or rods [118]. In one representative study of the beneficial effects of lactic acid bacteria, prebiotics (e.g. fructooligosaccharides) increased the growth of these microorganisms, which promoted the synthesis of lactic and butyric acids, and these two factors (especially the increasing lactic acid bacteria counts in the intestine) reduced colitis in rats [125]. In addition, it was demonstrated that strains of *L. acidophilus* produced heat stable antifungal metabolites with low pH optima [126]. The antipathogenic arsenal of lactic acid bacteria[§§§§] consists of the production of lactic acid and volatile acids, hydrogen peroxide and antibioticlike compounds, such as acidophilins or bacteriocins (e.g. nisin[*****]) [127-130]. As a final point, lactic acid bacteria are also known to release different enzymes into the intestinal lumen that exert synergistic effects on digestion, alleviating symptoms of intestinal malabsorption [128].

3.4 Butyric acid fermentation

Many microorganisms such as *Butyrivibrio fibrisolvens*, certain species of *Clostridium* (e.g. *C. acetobutylicum, C. butyricum, C. pasteurianum, C. perfringens, C. tyrobutyricum*) and *Fusobacterium nucleatum* carry out butyric acid fermentation [17, 131].

The end products of glucose fermentation include acetic and butyric acids **(Figure 15)**, CO_2 and molecular hydrogen (H_2) [17, 131, 132]. The 3-hydroxybutyryl-CoA dehydrogenase (EC 1.1.1.157) is the key enzyme in butyric acid production (its gene is clustered with those encoding other enzymes involved in butyric acid-butanol production on both the *C. acetobutylicum* and the *C. difficile* chromosomes) [133]. Despite that fact, microorganisms may present differences in metabolite production due to many factors.

[§§§§]Most probiotics fall into the group of lactic acid-producing bacteria and are normally consumed in the form of yogurts, fermented milks or other fermented foods – further discussed in the next chapter [128].
[*****]To date, nisin is the only bacteriocin that has found practical applications in some industrially processed foods [129].

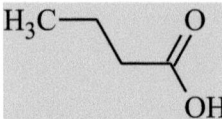

Figure 15:
Butyric acid: also known as butanoic acid, butanoate and butyrate, is a relevant short chain fatty acid whose production presents several potential beneficial effects on human health, as further discussed in the next chapter [132].

For instance, the production of butyric acid by *C. butyricum* ZJUCB at pH 6.5 was higher in fed-batch fermentation (16.74 g/L) than in batch fermentation (12.25 g/L) [134]. Acetone-butanol fermentation is carried out on an industrial scale to a limited extent by certain saccharolytic strains of *Clostridium* (e.g. *C. acetobutylicum*) in which glucose is initially metabolized via butyric acid fermentation, but subsequently the pH drops to ca. 4.5-5.0 and acetone and *n*-butanol are formed as major end products [17]. Other species (e.g. *C. perfringens*) form lactic acid and (or) ethanol in addition to the above products – lactic acid becoming a major end product under conditions of iron deficiency [17, 38]. The growth of anaerobic, sporeforming, lactate-fermenting butyric acid bacteria, especially of *C. tyrobutyricum*, may cause the late blowing of cheese due to excessive production of CO_2 and H_2 and a bad off-flavor [131]. Lately, much attention has been given to the functional character of butyric acid in human colonic fermentation. As previously mentioned, its pros and cons were discussed in subsection **(4.2)**.

3.5 Propionic acid fermentation

Generally, in propionic acid fermentation, glucose and (or) lactate yield propionic acid and acetic acid as the main end products (pyruvate is a common intermediate for both transformations) [17, 49]. Specifically, the first major pathway, the Wood-Werkman cycle, involves succinyl-CoA and methylmalonyl-CoA as intermediates and the second major pathway involves an acrylyl-CoA intermediate and has been described in *Clostridium propionicum* (see details further below) [135]. However, most propionibacteria produce propionic acid from pyruvate through the Wood-Werkman cycle (also known as a dicarboxylic acid cycle) – **Figure 16** [136].

Several microorganisms such as *Propionibacterium* spp., *Clostridium propionicum* and *Megasphaera elsdenii* carry out propionic acid fermentation [17]. *Propionibacterium* spp. ferment glucose or lactate via succinate, but when lactate is the substrate, the reduction of (endogenous) fumarate generates proton motive force[†††††], permitting ATP synthesis by electron transport phosphorylation [17, 137].

On the other hand, *C. propionicum*, *M. elsdenii* and *C. neopropionicum* X4 ferment lactate via a different pathway (the acrylate pathway) as schematically follows[‡‡‡‡‡]: glucose → pyruvate (1) → lactate (2) → lactyl-CoA (3) → acrylyl-CoA (4) → propionyl-CoA (5) → propionyl-P (6) → propionate [17, 138].

[†††††]The proton motive force is the chemical potential associated with moving a proton from the N-side to the P-side of the membrane and is sometimes termed a backpressure – briefly, N-side being the negative side of the membrane and P-side, the positive side of the membrane [137].

[‡‡‡‡‡]The numbers followed by arrows correspond to the enzymes and their products, respectively: (1) L-lactate dehydrogenase, (2) propionate CoA-transferase (EC 2.8.3.1), (3) lactoyl-CoA dehydratase (EC 4.2.1.54), (4) acyl-CoA dehydrogenase (EC 1.3.8.7, ambiguous), (5) phosphate acetyltransferase (EC 2.3.1.8), (6) propionate kinase (EC 2.7.2.15) [138].

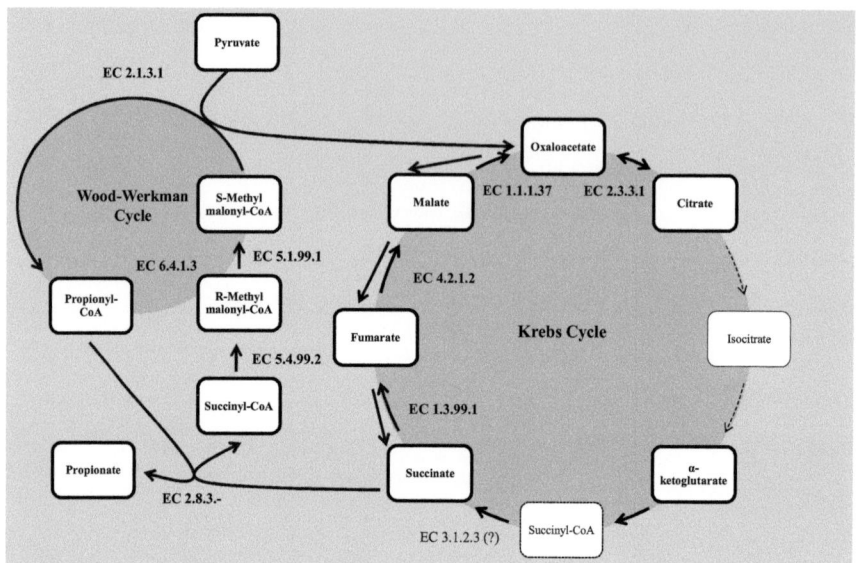

Figure 16:
Schematic representation of the Wood-Werkman cycle (adapted from [135, 139]).
Only the graphic elements in bold, *i.e.* metabolites, enzyme classes, lines and arrows indicate the metabolic flow of the Wood-Werkman cycle. It was first described in *Propionibacterium freudenreichii* and *Pelobacter propionicus*, but it is present in other bacterial species (e.g. *Bacteroïdes fragilis*, *Veillonella parvula* and *Veillonella gazogenes*) [135]. The key feature of the Wood-Werkman cycle involves the (biotin-dependent) enzyme methylmalonyl-CoA carboxytransferase (EC 2.1.3.1), transferring a carboxyl group from methylmalonyl-CoA to pyruvate to form oxaloacetate and propionyl-CoA [135]. It also includes a methylmalonyl-CoA epimerase (EC 5.1.99.1) and a methylmalonyl-CoA mutase (EC 5.4.99.2). A partial Krebs cycle was also reconstituted, but lacked a succinyl-CoA hydrolase (E.C.3.1.2.3) [135]. In aerobic organisms, the Krebs cycle is an amphibolic pathway, *i.e.* concerned in both anabolism (the energy-requiring part of metabolism in which simpler substances are transformed into more complex ones, as in growth and other biosynthetic processes) and catabolism (the breakdown of complex substances into smaller products, including the breakdown of carbon compounds with the liberation of energy for use by the cell or organism) [1, 49, 67].

Propionic acid bacteria are usually regarded as anaerobic or microaerophilic organisms, however, several *Propionibacterium* spp. possess the components of a typically aerobic electron transport chain, including membrane-bound dehydrogenases, menaquinones (subsection **4.2.3**) and cytochromes (actually, the capacity for oxidative phosphorylation has long been reported for *P. shermanii* and *P.*

peterssonii) [140]. They are Gram-positive, catalase positive, high G+C%[§§§§§], non-sporeforming and non-motile pleomorphic bacteria [141-144].

Propionic acid bacteria are only weakly lipolytic (e.g. via an intracellular esterase from *Propionibacterium freudenreichii* ssp. *freudenreichii* ITG 14) when compared with other groups of microorganisms, but due to the low taste threshold of some fatty acids, a large number of weakly lipolytic microorganisms may play a final important role in products that are stored for a long period such as ripened cheeses [145]. The metabolic activity of such economically important bacteria (e.g. the production of Swiss-type cheese) also leads to the characteristic eyes and nutty flavor in cheeses [146, 147]. These desirable characteristics can occur either spontaneously or be achieved by a culture of selected propionibacteria [146]. Diverse pathways are responsible for the formation of cheese flavor compounds: the Wood-Werkman cycle (for propionic acid formation), amino acid degradation pathways (for the formation of volatile branched chain fatty acids) and esterases (for the formation of free fatty acids and esters) [135]. **Figure 17.**

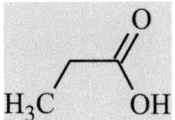

Figure 17:
Propionic acid (or propanoic acid).
It is commonly used as a mold inhibitor, but the antimicrobial capacity of propionic acid bacteria is only partly due to their production of organic acids; it has become evident that they also produce other biologically active substances (e.g. bacteriocins) [146].

[§§§§§]High G+C% is indicative of a higher neutral substitution rate, due to a higher rate of mutation at methylated cytosines in CpG sites (or islands) and some other factors [142, 143]. The abbreviation CpG stands for cytosine and guanine separated by a phosphate (—C—phosphate—G—), which links the two nucleosides together in DNA [144].

3.6 Succinic acid fermentation

Succinic acid, also known as butanedioic (or amber) acid, is a four-carbon dicarboxylic acid synthesized as an intermediate of the tricarboxylic acid cycle and also as one of the mixed-acid fermentation products [148]. For instance, the formation of mixed acids in *Mannheimia succiniciproducens* MBEL55E may be schematically represented as follows: glucose → phosphoenolpyruvate → oxalate → malate → fumarate → succinic acid [148]. Succinic acid is of high interest as bio-feedstock for the chemical industry since it is a precursor for a variety of many other chemicals, e.g. 1,4-butandiol, tetrahydrofuran, biodegradable polymers and fumaric acid [149]. It also functions as an acidulant and flavor enhancer in relishes, beverages and hot sausages [150]. However, up until now, data related to succinic acid fermentation remain very scarce, being only indicative of the potentiality of this fermentation pathway [149, 151-154]. **Figure 18**.

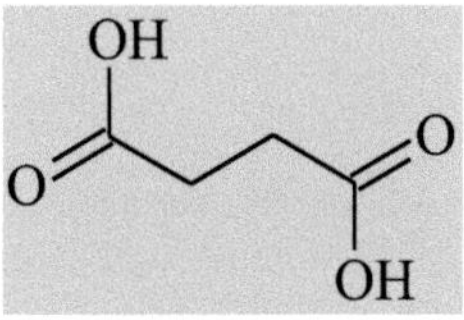

Figure 18:
Succinic acid (or butanedioic acid).
Succinic acid is also known as one of the major organic acids produced by yeast during fermentation for the production of alcoholic beverages [155]. Despite its relevance, the world market volume of succinic acid is relatively small at 15,000 tons per year [156].

3.7 An important role of Krebs cycle in microbial metabolic pathways

Besides the important role of Krebs cycle[******] in multicellular eukaryotes, it also plays a pivotal role in microbial metabolic pathways: it provides catabolic

---

[******]Sir Hans Adolf Krebs (1900-1981) [157].

pathways that arrive in its circuitry and anabolic pathways that leave from it in tight coordination with other metabolic pathways [1, 157]. The anabolic pathways include anaplerosis and cataplerosis. While anaplerosis replenishes the Krebs cycle intermediates that have been used for anabolic purposes[††††††], cataplerosis consumes the Krebs cycle intermediates for anabolic purposes too [49, 67, 158-160].

The most important anaplerotic reaction (in nature) is the conversion of pyruvate or phosphoenolpyruvate into oxaloacetate, and this transformation involves an essential carboxylation reaction that can be catalyzed by two different enzymes, phosphoenolopyruvate carboxylase (EC 4.1.1.31) and biotin-dependent pyruvate carboxylase (EC 6.4.1.1) – the latter being highly conserved in many organisms [160]. The Wood-Werkman cycle is an example of anaplerotic pathway specialized in the biosynthesis of propionic acid (see **Figure 16**) [161-163]. The glyoxylate shunt is also an anaplerotic pathway important for the metabolism of fatty acids, which is mediated by the enzyme isocitrate lyase (EC 4.1.3.1) along with malate synthase (EC 2.3.3.9) [164]. The glyoxylate cycle is very similar to the Krebs cycle, except for the reactions catalyzed by the two enzymes aforementioned [18].

As schematically represented above, succinic acid is synthesized from oxaloacetate, which is formed by anaplerotic reactions via malic acid and fumaric acid by the reductive Krebs cycle [165]. In view of that, it is important to know that under anaerobic conditions, the usual complete "aerobic" Krebs cycle is converted into two branches, an oxidative (or C6 branch) from citrate to 2-oxoglutarate and a reductive (or C4 branch) from oxaloacetate to succinate [166].

During biosynthetic processes, mitochondria act as regulators of the metabolic homeostasis in cells and regulate the carbon and nitrogen fluxes between proteins, lipids, by activating anaplerotic and cataplerotic reactions for reconstruction of membranes and supramolecular structures (e.g. ribosomes) [67, 167]. For instance, citrate cataplerosis from the Krebs cycle provides cytosolic acetyl-CoA for *de novo*

[††††††]Almost half of all amino acids are directly synthesized from Krebs cycle metabolites [160].

fatty acid synthesis and succinyl-CoA cataplerosis from the cycle provides carbon skeletons for the synthesis of heme groups [158]. These cataplerotic reactions of the Krebs cycle not only provide intermediates for the cellular biosynthesis, but when followed by anaplerotic reactions to regenerate oxaloacetate, also reduce the amount of NADH generated in the cycle [158]. To finish, it is worth remarking that these important metabolic aspects lack of pertinent studies.

REFERENCES

[1]   D. L. Nelson and M. M. Cox, *Lehninger Principles of Biochemistry*, W. H. Freeman and Company, New York, NY, USA, 2013.

[2]   F. R. X. Batista, K. N. Greco, R. M. Astray, S. A. C. Jorge, E. F. P. Augusto, C. A. Pereira, R. Z. Mendonça, and A. M. Moraes, "Behavior of wild-type and transfected S2 cells cultured in two different media," *Applied Biochemistry and Biotechnology*, vol. 163, no. 1, pp. 1-13, 2011.

[3]   L. A. Fothergill-Gilmore and P. A. M. Michels, "Evolution of glycolysis," *Progress in Biophysics and Molecular Biology*, vol, 59, no. 2, pp. 105-235, 1993.

[4]   S. Shanske, S. Sakoda, M. A. Hermodson, S. DiMauro, and E. A. Schon, "Isolation of a cDNA encoding the muscle-specific subunit of human phosphoglycerate mutase," *The Journal of Biological Chemistry*, vol. 262, no. 30, pp. 14612-14617, 1987.

[5]   S. Sakoda, S. Shanske, S. DiMauro, and E. A. Schon, "Isolation of a cDNA encoding the B isozyme of human phosphoglycerate mutase (PGAM) and characterization of the PGAM gene family," *The Journal of Biological Chemistry*, vol. 263, no. 32, pp. 16899-16905, 1988.

[6]   K. Yamada and T. Noguchi, "Nutrient and hormonal regulation of pyruvate kinase gene expression," *Biochemical Journal*, vol. 337, pp. 1-11, 1999.

[7]   B. Canback, S. G. E. Andersson, and C. G. Kurland, "The global phylogeny of

glycolytic enzymes," *Proceedings of the National Academy of Sciences*, vol. 99, no. 9, pp. 6097-6102, 2002.

[8]   Y. Zhao, R. E. Davis, and I.-M. Lee, "Phylogenetic positions of '*Candidatus* Phytoplasma asteris' and *Spiroplasma kunkelii* as inferred from multiple sets of concatenated core housekeeping proteins," *International Journal of Systematic and Evolutionary Microbiology*, vol. 55, no. 5, pp. 2131-2141, 2005.

[9]   J. Cho, J. S. King, X. Qian, A. J. Harwood, and S. B. Shears, "Dephosphorylation of 2,3-bisphosphoglycerate by MIPP expands the regulatory capacity of the Rapoport–Luebering glycolytic shunt," *Proceedings of the National Academy of Sciences*, vol. 105, no. 16, pp. 5998-6003, 2008.

[10]   R. A. Harris, "Carbohydrate metabolism I: major metabolic pathways and their control," in *Text Book of Biochemistry with Clinical Correlation*, T. M. Develin, Ed., pp. 267-333, Wiley-Liss, New York, NY, USA, 4th edition, 1997.

[11]   K. E. Conley, M. J. Kushmerick, and S. A. Jubrias, "Glycolysis is independent of oxygenation state in stimulated human skeletal muscle *in vivo*," *Journal of Physiology*, vol. 511, no. 3, pp. 935-945, 1998.

[12]   B. A. Maron, B. A. Cockrill, A. B. Waxman, and D. M. Systrom, "The invasive cardiopulmonary exercise test," *Circulation*, vol. 127, pp. 1157-1164, 2013.

[13]   M. F. Bachmann, "Taurine: energy drink for T cells," *European Journal of Immunology*, vol. 42, no. 4, pp. 819-821, 2012.

[14]   L. A. Silva, M. C. Vinaud, A. M. Castro, P. V. L. Cravo, and J. C. B. Bezerra, "*In silico* search of energy metabolism inhibitors for alternative Leishmaniasis treatments," *BioMed Research International*, vol. 2015, Article ID 965725, 2015.

[15]   A. Leone, E. Di Gennaro, F. Bruzzese, A. Avallone, and A. Budillon, "New perspective for an old antidiabetic drug: metformin as anticancer agent," *Advances in Nutrition and Cancer*, vol. 159, pp 355-376, 2014.

[16]   T. E. Graham, "Measurement and interpretation of lactate," in *Progress in Ergometry: Quality Control and Test Criteria*, H. Löllgen and H. Mellerowicz, Eds.,

pp 51-66, Springer, Berlin, Germany, 1984.

[17] P. Singleton and D. Sainsbury, *Dictionary of Microbiology and Molecular Biology*, John Wiley & Sons, London, UK, 2006.

[18] J. L. Tymoczko, J. M. Berg, and L. Stryer, *Biochemistry*, Worth Publishers, New York, NY, USA, 2010.

[19] B. D. Prasad, G. Thapa, S. Baishya, and S. Sahni, "Biochemical and molecular characterization of submergence tolerance in rice for crop improvement," *Journal of Plant Breeding and Crop Science*, vol. 3, no. 10, pp. 240-250, 2011.

[20] C. W. Bamforth, *Food, Fermentation and Micro-organisms*, Blackwell Publishing, Ames, IA, USA, 2005.

[21] *Churchill's Medical Dictionary*, E. L. Becker, Lord Butterfield of Stechford, A. M. Harvey, R. H. Heptinstall, and L. Thomas, Eds., Churchill Livingstone, New York, NY, USA, 1989.

[22] *Dorland's Illustrated Medical Dictionary*, D. A. Anderson, J. M. Patwell, K. Plaut, and K. McCullough, Eds., W. B. Saunders Company, Philadelphia, PA, USA, 1988.

[23] F. Zamora, "Biochemistry of alcoholic fermentation," in *Wine Chemistry and Biochemistry*, M. V. Moreno-Arribas and M. C. Polo, Eds., pp. 3-26, Springer, New York, NY, USA, 2009.

[24] Z. Berk, *Food Process Engineering and Technology*, Elsevier, New York, NY, USA, 2009.

[25] A. Skelin, S. Sikora, S. Orlić, L. Duraković, and S. Redžepović, "Genetic diversity of indigenous *Saccharomyces sensu stricto* yeasts isolated from Southern Croatia," *Agriculturae Conspectus Scientificus*, vol. 73, no. 2, pp. 89-94, 2008.

[26] R. Olmi, V. V. Meriakri, A. Ignesti, S. Priori, and C. Riminesi, "Monitoring alcoholic fermentation by microwave dielectric spectroscopy," *Journal of Microwave Power & Electromagnetic Energy*, vol. 41, no. 3, pp. 38-50, 2007.

[27] A. N. Emmanuel, A. David, Y. K. Benjamin, and E. T. Yannick, "A hybrid

neural network approach for batch fermentation simulation," *Australian Journal of Basic and Applied Sciences*, vol. 3, no. 4, pp. 3930-3936, 2009.

[28]   M. Begea, M. Vlădescu, G. Bâldea, C. Cîmpeanu, C. Stoicescu, and P. Begea, "Isolation and selection of high ethanol producing yeast strains," *Journal of Agroalimentary Processes and Technologies*, vol. 15, no. 1, pp. 107-111, 2009.

[29]   N. T Bendsen, R. Christensen, E. M. Bartels, F. J. Kok, A. Sierksma, A. Raben, and A. Astrup, "Is beer consumption related to measures of abdominal and general obesity? A systematic review and meta-analysis," *Nutrition Reviews*, vol. 71, no. 2, pp. 67-87, 2013.

[30]   D. Grant and J. A. Lewis, "Zero tolerance," in *The Encyclopedia of Theoretical Criminology*, J. M. Miller, Ed., pp. 1-4, John Wiley & Sons, Ltd, 2014.

[31]   J. A. Diderich, L. M. Raamsdonk, A. L. Kruckeberg, J. A. Berden, and K. V. DAM, "Physiological properties of *Saccharomyces cerevisiae* from which hexokinase II has been deleted," *Applied and Environmental Microbiology*, vol. 67, no. 4, pp. 1587-1593, 2001.

[32]   S. Hohmann, "Characterization of PDC6, a third structural gene for pyruvate decarboxylase in *Saccharomyces cerevisiae*," *Journal of Bacteriology*, vol. 173, no. 24, pp. 7963-7969, 1991.

[33]   J. T. Pronk, H. Y. Steensma, and J. P. van Dijken, "Pyruvate metabolism in *Saccharomyces cerevisiae*," *Yeast*, vol. 12, no. 16, pp. 1607-1633, 1996.

[34]   M, Tadege, I. Dupuis, and C. Kuhlemeier, "Ethanolic fermentation: new functions for an old pathway," *Trends in Plant Science*, vol. 4, no. 8, pp. 320-325, 1999.

[35]   C. W. Chan, K. L. Cruz, H. Kalra, B. Tam, and C. Yang, "Investigation of cell density of BY4741A wild-type and YLR044C mutant strain *Saccharomyces cerevisiae* in response to varying dextrose conditions," *The Expedition*, vol. 3, 2013.

[36]   E, Skovran and D. M. Downs, "Metabolic defects caused by mutations in the isc gene cluster in *Salmonella enterica* serovar Typhimurium: implications for

thiamine synthesis," *Journal of Bacteriology*, vol. 182, no. 14, pp. 3896-3903, 2000.

[37] J. H. Kirkels, C. J. A. van Echteld, and T. J. C. Ruigrok, "Intracellular magnesium during myocardial ischemia and reperfusion: possible consequences for postischemic recovery," *Journal of Molecular and Cellular Cardiology*, vol. 21, no. 11, pp. 1209-1218, 1989.

[38] K. H. Steinkraus, "Classification of fermented foods: worldwide review of household fermentation techniques," *Food Control*, vol. 8, no. 5/6, pp. 311-317, 1997.

[39] P. Chumnanpuen, M. A. E. Hansen, J. Smedsgaard, and J. Nielsen, "Dynamic metabolic footprinting reveals the key components of metabolic network in yeast *Saccharomyces cerevisiae*," *International Journal of Genomics*, vol. 2014, Article ID 894296, 2014.

[40] F. Alam, Md. A. Islam, S. H. Gan, and Md. I. Khalil, "Honey: a potential therapeutic agent for managing diabetic wounds," *Evidence-Based Complementary and Alternative Medicine*, vol. 2014, Article ID 169130, 2014.

[41] P. M. Christy, L. R. Gopinath, and D. Divya, "A review on anaerobic decomposition and enhancement of biogas production through enzymes and microorganisms," *Renewable and Sustainable Energy Reviews*, vol. 34, 167-173, 2014.

[42] S. Reiter, "Barriers to effective treatment of vaginal atrophy with local estrogen therapy," *International Journal of General Medicine*, vol. 6, pp. 153-158, 2013.

[43] M. Ambye-Jensen, S. T. Thomsen, Z. Kádár, and A. S. Meyer, "Ensiling of wheat straw decreases the required temperature in hydrothermal pretreatment," *Biotechnology for Biofuels*, vol. 6, no. 116, 2013.

[44] C. Larsson, A. Nilsson, A. Blomberg, and L. Gustafsson, "Glycolytic flux is conditionally correlated with ATP concentration in *Saccharomyces cerevisiae*: a chemostat study under carbon- or nitrogen-limiting conditions," *Journal of*

Bacteriology, vol. 179, no. 23, pp. 7243-7250, 1997.

[45] Y. Ida, C. Furusawa, T. Hirasawa, and H. Shimizu, "Stable disruption of ethanol production by deletion of the genes encoding alcohol dehydrogenase isozymes in *Saccharomyces cerevisiae*," *Journal of Bioscience and Bioengineering*, vol. 113, no. 2, pp. 192-195, 2012.

[46] A. Gauf, C. Navarro, G. Balch, L. R. Hargreaves, and M. A. Khakoo, "Low-energy elastic electron scattering by acetaldehyde," *Physical Review A*, vol. 89, pp. 022708-1-022708-8, 2014.

[47] G. Xu, Q. Hua, N. Duan, L. Liu, and J. Chen, "Regulation of thiamine synthesis in *Saccharomyces cerevisiae* for improved pyruvate production," *Yeast*, vol. 29, pp. 209-217, 2012.

[48] C. Brion, C. Ambroset, P. Delobel, I. Sanchez, and B. Blondin, "Deciphering regulatory variation of THI genes in alcoholic fermentation indicate an impact of Thi3p on *PDC1* expression," *BMC Genomics*, 15:1085, 2014.

[49] *Oxford Dictionary of Biochemistry and Molecular Biology*, R. Cammack, Ed., Oxford University Press Inc., New York, NY, USA, 2006.

[50] V. M. Anoop, U. Basu, M. T. McCammon, L. McAlister-Henn, and G. J. Taylor, "Modulation of citrate metabolism alters aluminum tolerance in yeast and transgenic canola overexpressing a mitochondrial citrate synthase," *Plant Physiology*, vol. 132, no. 4, pp. 2205-2217, 2003.

[51] E. Toro and L. Shapiro, "Bacterial chromosome organization and segregation," in *Additional Perspectives on Cell Biology of Bacteria*, L. Shapiro and R. Losick, Eds., pp. 1-15, Cold Spring Harbor Laboratory Press, 2:a000349, 2010.

[52] C. A. Lewis, S. J. Parker, B. P. Fiske, D. McCloskey, D. Y. Gui, C. R. Green, N. I. Vokes, A. M. Feist, M. G. V. Heiden, and C. M. Metallo, "Tracing compartmentalized NADPH metabolism in the cytosol and mitochondria of mammalian cells," *Molecular Cell*, vol. 55, no. 2, pp. 253-263, 2014.

[53] C. Zimmer and E. Fabre, "Principles of chromosomal organization: lessons

from yeast," *The Journal of Cell Biology*, vol. 192, no. 5, pp. 723-733, 2011.

[54] J. Perdigão and C. E. Sunkel, "The centromere: structural and functional aspects," *Advances in Genome Biology*, vol. 5B, pp. 263-321, 1998.

[55] G. Marx, A. Moody, and D. Bermúdez-Aguirre, "A comparative study on the structure of *Saccharomyces cerevisiae* under nonthermal technologies: high hydrostatic pressure, pulsed electric fields and thermo-sonication," *International Journal of Food Microbiology*, vol. 151, no.3, pp. 327-337, 2011.

[56] T. Hirasawa, Y. Ida, C. Furusawa, and H. Shimizu, "Potential of a *Saccharomyces cerevisiae* recombinant strain lacking ethanol and glycerol biosynthesis pathways in efficient anaerobic bioproduction," *Bioengineered*, vol. 5, no. 2, pp. 123-128, 2014.

[57] F. Rolland, J. Winderickx, and J. M. Thevelein, "Glucose-sensing mechanisms in eukaryotic cells," *TRENDS in Biochemical Sciences*, vol. 26, no. 5, pp. 310- 317, 2001.

[58] S. Pasula, S. Chakraborty, J. H. Choi, and J.-H. Kim, "Role of casein kinase 1 in the glucose sensor-mediated signaling pathway in yeast," *BMC Cell Biology*, vol. 11, no. 17, 2010.

[59] E. Simeonidis, E. Murabito, K. Smallbone, and H. V. Westerhof, "Why does yeast ferment? A flux balance analysis study," *Biochemical Society Transactions*, vol. 38, no. 5, pp. 1225-1229, 2010.

[60] Y. S. Jin, H. Ni, J. M. Laplaza, and T. W. Jeffries, " Optimal growth and ethanol production from xylose by recombinant *Saccharomyces cerevisiae* require moderate D-xylulokinase activity," *Applied and Environmental Microbiology*, vol. 69, no. 1, pp. 495-503, 2003.

[61] S. Fernandes and P. Murray, "Metabolic engineering for improved microbial pentose fermentation," *Bioengineered Bugs*, vol. 1, no. 6, pp. 424-428, 2010.

[62] M. H. Toivari, L. Salusjärvi, L. Ruohonen, and M. Penttila, "Endogenous xylose pathway in *Saccharomyces cerevisiae*," *Applied and Environmental*

Microbiology, vol. 70, no. 6 pp. 3681-3686, 2004.

[63] K. L. Träff, R. R. O. Cordero, W. H. van Zyl, and B. Hahn-Hägerdal, "Deletion of the GRE3 aldose reductase gene and its influence on xylose metabolism in recombinant strains of *Saccharomyces cerevisiae* expressing the xylA and XKS1 genes," *Applied and Environmental Microbiology*, vol. 67, no. 12, pp. 5668-5674, 2001.

[64] G. I. Naumov, E. S. Naumova, N. N. Martynenko, and I. Masneuf-Pomarédec, "Taxonomy, ecology, and genetics of the yeast *Saccharomyces bayanus*: a new object for science and practice," *Microbiology*, vol. 80, no. 6, pp. 735-742, 2011.

[65] R. W. Hutkins, *Microbiology and Technology of Fermented Foods*, Blackwell Publishing, Ames, IA, USA, 2006.

[66] S. Rainieri, "The brewer's yeast genome: from its origins to our current knowledge," in *Beer in Health and Disease Prevention*, V. R. Preedy, Ed., pp. 89-101, Elsevier, London, NW, UK, 2009.

[67] P.-S. Juo, *Concise Dictionary of Biomedicine and Molecular Biology*, CRC Press, Boca Raton, FL, USA, 2002.

[68] E. Zörgö, A. Gjuvsland, F. A. Cubillos, E. J. Louis, G. Liti, A. Blomberg, S. W. Omholt, and Jonas Warringer, "Life history shapes trait heredity by accumulation of loss-of-function alleles in yeast," *Molecular Biology and Evolution*, vol. 29, no. 7, pp. 1781-1789, 2012.

[69] F. M. P. G. Ernandes and C. H. Garcia-Cruz, "*Zymomonas mobilis*: a promising microrganism for alcoholic fermentation," *Semina: Ciências Agrárias*, vol. 30, no. 2, pp. 361-380, 2009.

[70] V. K. Joshi, S. K. Chauhan, and S. Bhushan, "Technology of fruit-based alcoholic beverages," in *Postharvest Technology of Fruits and Vegetables: : Handling, Processing, Fermentation and Waste Management*, Volume 2: Technology, L. R. Verma and V. K. Joshi, Eds., Indus Publishing Co., New Delhi, India, 2000.

[71] Z. Genisheva, S. I. Mussatto, J. M. Oliveira, and J. A. Teixeira, "Malolactic fermentation of wines with immobilised lactic acid bacteria – influence of concentration, type of support material and storage conditions," *Food Chemistry*, vol. 138, no. 2-3, pp. 1510-1514, 2013.

[72] R. Rodicio and J. J. Heinisch, "Yeast on the milky way: genetics, physiology and biotechnology of *Kluyveromyces lactis*," *Yeast*, vol. 30, no. 5, pp. 165-177, 2013.

[73] I. Orlandi, R. Ronzulli, N. Casatta, and M. Vai, "Ethanol and acetate acting as carbon/energy sources negatively affect yeast chronological aging," *Oxidative Medicine and Cellular Longevity*, vol. 2013, Article ID 802870, 2013.

[74] B. J. Giardina, B. A. Stanley, and H.-L. Chiang, "Glucose induces rapid changes in the secretome of *Saccharomyces cerevisiae*," *Proteome Science*, vol. 12, no. 9, 2014.

[75] G. Feiner, *Meat Products Handbook: Practical Science and Technology*, CRC Press LLC, Boca Raton, FL, USA, 2006.

[76] B. L. Tang, "The mitochondrial pyruvate carrier and metabolic regulation," *CellBio*, vol. 3, no. 4, pp. 111-117, 2014.

[77] A. Kokavec, "Decreased appetite for food in alcoholism," in *Handbook of Behavior, Food and Nutrition*, V. R. Preedy, R. R. Watson, and C. R. Martin, Eds., pp. 2949-2962, Springer Science+Business Media, LLC, New York, NY, USA, 2011.

[78] N. Ishida, S. Saitoh, K. Tokuhiro, E. Nagamori, T. Matsuyama, K. Kitamoto, and H. Takahashi, "Efficient production of L-lactic acid by metabolically engineered *Saccharomyces cerevisiae* with a genome-integrated L-lactate dehydrogenase gene," *Applied and Environmental Microbiology*, vol. 71, no. 4, pp. 1964-1970, 2005.

[79] S. Motomura, K. Horie, and H. Kitagaki, "Mitochondrial activity of sake brewery yeast affects malic and succinic acid production during alcoholic fermentation," *Journal of the Institute of Brewing*, vol. 118, no. 1, pp. 22-26, 2012.

[80] J. A. Barnett and K.-D. Entian, "A history of research on yeasts 9: regulation of sugar metabolism," *Yeast*, vol. 22, no. 11, pp. 835-894, 2005.

[81]   A. Hagman, T. Säll, and J. Piškur, "Analysis of the yeast short-term Crabtree effect and its origin," *The FEBS Journal*, vol. 281, no. 21, pp. 4805-4814, 2014.

[82]   S. Dashko, N. Zhou, C. Compagno, and J. Piškur, "Why, when, and how did yeast evolve alcoholic fermentation?," *FEMS Yeast Research*, vol. 14, no. 6, pp. 826-832, 2014.

[83]   M. R. Wijsman, J. P. van Dijken, B. H. A. van Kleeff, and W. A. Scheffers, "Inhibition of fermentation and growth in batch cultures of the yeast *Brettanomyces intermedius* upon a shift from aerobic to anaerobic conditions (Custers effect)," *Antonie van Leeuwenhoek*, vol. 50, no. 2, pp 183-192, 1984.

[84]   A. Morata, R. Vejarano, G. Ridolfi, S. Benito, F. Palomero, C. Uthurry, W. Tesfaye, C. González, and J. A. Suárez-Lepe, "Reduction of 4-ethylphenol production in red wines using HCDC+ yeasts and cinnamyl esterases," *Enzyme and Microbial Technology*, vol. 52, no. 5, pp. 99-104, 2013.

[85]   J.-X. Fontaine, T. Tercé-Laforgue, S. Bouton, K. Pageau, P. J. Lea, F. Dubois, and B. Hirel, "Further insights into the isoenzyme composition and activity of glutamate dehydrogenase in *Arabidopsis thaliana*," *Plant Signaling & Behavior*, vol. 8, no. 3, pp. e23329-1-e23329-5, 2013.

[86]   T. C. James, S. Campbell, D. Donnelly, and U. Bond, "Transcription profile of brewery yeast under fermentation conditions," *Journal of Applied Microbiology*, vol. 94, no. 3, pp. 432-448, 2003.

[87]   O. de Smidt, J. C. du Preez, and J. Albertyn, "The alcohol dehydrogenases of *Saccharomyces cerevisiae*: a comprehensive review," *FEMS Yeast Research*, vol. 8, no. 7, pp. 967-978, 2008.

[88]   W. A. Wilson, P. J. Roach, M. Montero, E. Baroja-Fernández, F. J. Muñoz, G. Eydallin, A. M. Viale, and J. Pozueta-Romero, "Regulation of glycogen metabolism in yeast and bacteria," *FEMS Microbiology Reviews*, vol. 34, no. 6, pp. 952-985, 2010.

[89]   I. Eliaz, "An integrative approach to enhancing cognition and executive brain

function: a personal perspective," *Archives of Agronomy and Soil Science*, vol. 20, no. 1, 27-30, 2014.

[90] J. Gebauer, S. Schuster, L. F. de Figueiredo, and C. Kaleta, "Detecting and investigating substrate cycles in a genome-scale human metabolic network," *The FEBS Journal*, vol. 279, no. 17, pp. 3192-3202, 2012.

[91] L. F. de Figueiredo, T. I. Gossmann, M. Ziegler, S. Schuster, "Pathway analysis of NAD+ metabolism," *Biochemical Journal*, vol. 439, no. 2, pp.341-348, 2011.

[92] G. Kim, S. J. Weiss, and R. L. Levine, "Methionine oxidation and reduction in proteins," *Biochimica et Biophysica Acta (BBA) – General Subjects*, vol. 1840, no. 2, pp. 901-905, 2014.

[93] M. V. Semkiv, K. V. Dmytruk, C. A. Abbas, and A. A. Sibirny, "Increased ethanol accumulation from glucose via reduction of ATP level in a recombinant strain of *Saccharomyces cerevisiae* overexpressing alkaline phosphatase," *BMC Biotechnology*, vol. 14, no. 42, 2014.

[94] S. de Kok, B. U. Kozak, J. T. Pronk, and A. J. A. van Maris, "Energy coupling in *Saccharomyces cerevisiae*: selected opportunities for metabolic engineering," *FEMS Yeast Research*, vol. 12, no. 4, pp. 387-397, 2012.

[95] J. M. Ferreira, R. Swarnakar, and F. L. H. Silva, "Effect of nutrient sources on bench scale vinegar production using response surface methodology," *Revista Brasileira de Engenharia Agrícola e Ambiental*, vol. 9, no. 1, pp. 73-77, 2005.

[96] W. Tesfaye, M. L. Morales, M. C. García-Parrilla, and A. M. Troncoso, "Wine vinegar: technology, authenticity and quality evaluation," *Trends in Food Science & Technology*, vol. 13, no. 1, pp. 12-21, 2002.

[97] J. C. Yoo, J. B. Sim, H. K. Kim, H. S. Chun, and S. J. Kim, "Purification and properties of a membrane-bound alcohol dehydrogenase from *Acetobacter* sp. HA," *The Korean Journal of Microbiology*, vol. 32, 1, pp. 78-83, 1994.

[98] L. Lumeng and E. J. Davis, "Mechanism of ethanol suppression of

gluconeogenesis," *The Journal of Biological Chemistry*, vol. 245, no. 12, pp. 3179-3185, 1970.

[99] A. Iida, Y. Ohnishi, and S. Horinouchi, "Control of acetic acid fermentation by quorum sensing via *N*-acylhomoserine lactones in *Gluconacetobacter intermedius*," *Journal of Bacteriology*, vol. 190, no. 7, pp. 2546-2555, 2008.

[100] C. B. Rasrendra, B. Girisuta, H. H. van de Bovenkamp, J. G. M. Winkelman, E. J. Leijenhorst, R. H. Venderbosch, M. Windt, D. Meier, and H. J. Heeresa, "Recovery of acetic acid from an aqueous pyrolysis oil phase by reactive extraction using tri-*n*-octylamine," *Chemical Engineering Journal*, vol. 176-177, pp. 244-252, 2011.

[101] N. P. Mira, S. F. Henriques, G. Keller, M. C. Teixeira, R. G. Matos, C. M. Arraiano, D. R. Winge, and I. Sá-Correia, "Identification of a DNA-binding site for the transcription factor Haa1, required for *Saccharomyces cerevisiae* response to acetic acid stress," *Nucleic Acids Research*, vol. 39, no. 16, pp. 6896-6907, 2011.

[102] N. Wei, J. Quarterman, S. R. Kim, J. H. D. Cate, and Y.-S. Jin, "Enhanced biofuel production through coupled acetic acid and xylose consumption by engineered yeast," *Nature Communications*, vol. 4, no. 2580, 2013.

[103] K. B. Maal, R. Shafiei, and N. Kabiri, "Production of apricot vinegar using an isolated *Acetobacter* strain from Iranian apricot," *World Academy of Science, Engineering and Technology*, vol. 4, no. 11, pp. 162-165, 2010.

[104] I. Y. Sengun and S. Karabiyikli, "Importance of acetic acid bacteria in food industry," *Food Control*, vol. 22, no. 5, pp. 647-656, 2011.

[105] E. J. Bartowsky and P. A. Henschke, "Acetic acid bacteria spoilage of bottled red wine — a review," *International Journal of Food Microbiology*, vol. 125, no. 1. pp. 60-70, 2008.

[106] E. S. Bulygina, O. M. Gulikova, E. M. Dikanskaya, A. I. Netrusov, T. P. Tourova, and K. M. Chumakov, "Taxonomic studies of the genera *Acidomonas*, *Acetobacter* and *Gluconobacter* by 5S ribosomal RNA sequencing," *Journal of*

General Microbiology, vol. 138, no. 11, pp. 2283-2286, 1992.

[107] W. J. du Toit and I. S. Pretorius, "The occurrence, control and esoteric effect of acetic acid bacteria in winemaking," *Annals of Microbiology*, vol. 52, pp. 155-179, 2002.

[108] H. Jain and S. Mulay, "A review on different modes and methods for yielding a pentose sugar: xylitol," *International Journal of Food Sciences and Nutrition*, vol. 65, no. 2, pp. 135-143, 2014.

[109] S. M. M. Kamal, N. L. Mohamad, A. G. L. Abdullah, and N. Abdullah, "Detoxification of sago trunk hydrolysate using activated charcoal for xylitol production," *Procedia Food Science*, vol. 1, pp. 908-913, 2011.

[110] M. Setorki, S. Asgary, S. Haghjooyjavanmard, and B. Nazari, "Reduces cholesterol induced atherosclerotic lesions in aorta artery in hypercholesterolemic rabbits," *Journal of Medicinal Plants Research*, vol. 5, no. 9, pp. 1518-1525, 2011.

[111] G. Molin, "Probiotics in foods not containing milk or milk constituents, with special reference to *Lactobacillus plantarum* 299v," *The American Journal of Clinical Nutrition*, vol. 73, no. 2, pp.380s-385s, 2001.

[112] S. Tanasupawat and W. Visessanguan, "Fish fermentation," in *Seafood Processing: Technology, Quality and Safety*, I. S. Boziaris, Ed., pp. 178-207, John Wiley & Sons, Ltd, Chichester, UK, 2014.

[113] J. H. Chen, Y. Ren, J. Seow, T. Liu, W. S. Bang, and H. G. Yuk, "Intervention technologies for ensuring microbiological safety of meat: current and future trends," *Comprehensive Reviews in Food Science and Food Safety*, vol. 11, no. 2, pp. 119-132, 2012.

[114] S. Kumar, N. Prakash, and D. Datta, "Biopolymers based on carboxylic acids derived from renewable resources," in *Biopolymers: Biomedical and Environmental Applications*, S. Kalia and L. Avérous, Eds., pp. 169-182, Scrivener Publishing LLC, Salem, MA, USA, 2011.

[115] F. A. C. Martinez, E. M. Balciunas, J. M. Salgado, J. M. D. González, A.

Converti, and R. P. de S. Oliveira, "Lactic acid properties, applications and production: a review," *Trends in Food Science & Technology*, vol. 30, no. 1, pp. 70-83, 2013.

[116] S. Liang, A. G. McDonald, and E. R. Coats, "Lactic acid production with undefined mixed culture fermentation of potato peel waste," *Waste Management*, vol. 34, no. 11, pp. 2022-2027, 2014.

[117] M. H. Thomsen, J. P. Guyot, and P. Kiel, "Batch fermentations on synthetic mixed sugar and starch medium with amylolytic lactic acid bacteria," *Applied Microbiology and Biotechnology*, vol. 74, no. 3, pp. 540-546, 2007.

[118] A. Antunes, F. A. Rainey, M. F. Nobre, P. Schumann, A. M. Ferreira, A. Ramos, H. Santos, and M. S. da Costa, "*Leuconostoc ficulneum* sp. nov., a novel lactic acid bacterium isolated from a ripe fig, and reclassification of *Lactobacillus fructosus* as *Leuconostoc fructosum* comb. nov.," *International Journal of Systematic and Evolutionary Microbiology*, vol. 52, no. 2, pp. 647-655, 2002.

[119] Y. Le Loir, V. Azevedo, S. C. Oliveira, D. A. Freitas, A. Miyoshi, L. G. Bermúdez-Humarán, S. Nouaille, L. A. Ribeiro, S. Leclercq, J. E. Gabriel, V. D. Guimaraes, M. N. Oliveira, C. Charlier, M. Gautier, and P. Langella, "Protein secretion in *Lactococcus lactis*: an efficient way to increase the overall heterologous protein production," *Microbial Cell Factories*, vol. 4, no. 2, 2005.

[120] D. A. Freitas, S. Leclerc, A. Miyoshi, S. C. Oliveira, P. S. M. Sommer, L. Rodrigues, A. Correa Junior, M. Gautier, P. Langella, V. A. Azevedo, and Y. Le Loir, "Secretion of *Streptomyces tendae* antifungal protein 1 by *Lactococcus lactis*," *Brazilian Journal of Medical and Biological Research*, vol. 38, no. 11, pp. 1585-1592, 2005.

[121] A. Kumari, A. P. Garg, K. Makeen, M. Lal, C. Gupta, and S. Chandra, "A bacteriocin production on soya nutri nuggets extract medium by *Lactococcus lactis* subsp. *lactis* CCSUB202," *International Journal of Dairy Sci*ence, vol. 3, no. 1, pp. 49-54, 2008.

[122] L. Tserovska, S. Stefanova, and T. Yordanova, "Identification of lactic acid bacteria isolated from katyk, goat's milk and cheese," *Journal of Culture Collections*, vol. 3, pp. 48-52, 2000-2002.

[123] R. Priya, B. Bharathi, A. Jayanthi, and R. Vidhyalakshmi, "Studies on isolation, characterization and entrapment of probiotic *Lactobacillus* from milk and milk products," *Journal of Advanced Biotechnology*, vol. 10, no. 9, pp. 30-33, 2011.

[124] J. Magnusson, H. Jonsson, J. Schnürer, and S. Roos, "*Weissella soli* sp. nov., a lactic acid bacterium isolated from soil," *International Journal of Systematic and Evolutionary Microbiology*, vol. 52, no. 3, pp. 831-834, 2002.

[125] C. Cherbut, C. Michel, and G. Lecannu, "The prebiotic characteristics of fructooligosaccharides are necessary for reduction of TNBS-induced colitis in rats," *Journal of Nutrition*, vol. 133, no. 1, pp. 21-27, 2003.

[126] C. de Muyncka, A. I. J. Leroya, S. de Maeseneirea, F. Arnautb, W. Soetaerta, and E. J. Vandammea, "Potential of selected lactic acid bacteria to produce food compatible antifungal metabolites," *Microbiological Research*, vol. 159, no. 4, 339-346, 2004.

[127] S. J. Bhatia, N. Kochar, P. Abraham, N. G. Nair, and A. P. Mehta, "*Lactobacillus acidophilus* inhibits growth of *Campylobacter pylori* in vitro," *Journal of Clinical Microbiology*, vol. 27, no. 10, pp. 2328-2330, 1989.

[128] S. Parvez, K. A. Malik, S. Ah Kang, and H. Y. Kim, "Probiotics and their fermented food products are beneficial for health," *Journal of Applied Microbiology*, vol. 100, no. 6, pp. 1171-1185, 2006.

[129] C. Gupta, D. Prakash, and S. Gupta, "Natural useful therapeutic products from microbes," *Journal of Microbiology & Experimentation*, vol. 1, no. 1, 2014.

[130] T. Kawarai, S. Furukawa, H. Ogihara, and M. Yamasaki, "Mixed-species biofilm formation by lactic acid bacteria and rice wine yeasts," *Applied and Environmental Microbiology*, vol. 73, no. 14, pp. 4673-4676, 2007.

[131] G. van den Berg, W.C. Meijer, E. M. Düsterhöft, and G. Smit, "Gouda and

related cheeses," in *Cheese: Chemistry, Physics and Microbiology*, P. F. Fox, P. L. H. McSweeney, T. M. Cogan, and T. P. Guinee, Eds., pp. 103-140, Elsevier Academic Press, Oxford, UK, 3rd edition, 2004.

[132] C. J. Meehan and R. G. Beiko, "A phylogenomic view of ecological specialization in the Lachnospiraceae, a family of digestive tract-associated bacteria," *Genome Biology and Evolution*, vol. 6, no. 3, pp. 703-713, 2014.

[133] S. Karlsson, B. Dupuy, K. Mukherjee, E. Norin, L. G. Burman, and T. Åkerlund, "Expression of *Clostridium difficile* toxins A and B and their sigma factor TCDD is controlled by temperature," *Infection and Immunity*, vol. 71, no. 4, pp. 1784-1793, 2003.

[134] G. He, Q. Kong, Q. Chen, and H. Ruan, "Batch and fed-batch production of butyric acid by *Clostridium butyricum* ZJUCB," *Journal of Zhejiang University Science B*, vol. 6, no. 11, pp. 1076-1080, 2005.

[135] H. Falentin, S.-M. Deutsch, G. Jan, V. Loux, A. Thierry, S. Parayre, M.-B. Maillard, J. Dherbécourt, F. J. Cousin, J. Jardin, P. Siguier, A. Couloux, V. Barbe, B. Vacherie, P. Wincker, J.-F. Gibrat, C. Gaillardin, and S. Lortal, "The complete genome of *Propionibacterium freudenreichii* CIRM-BIA1[T], a hardy *Actinobacterium* with food and probiotic applications," *PLoS ONE*, vol. 5, no. 7, e11748, 2010.

[136] C. C. Stowers, B. M. Cox, and B. A. Rodriguez, "Development of an industrializable fermentation process for propionic acid production," *Journal of Industrial Microbiology & Biotechnology*, vol. 41, no. 5, pp. 837-852, 2014.

[137] J. N. Bazil, K. C. Vinnakota, F. Wu, and D. A. Beard, "Analysis of the kinetics and bistability of ubiquinol: cytochrome *c* oxidoreductase," *Biophysical Journal*, vol. 105, no. 2, pp. 343-355, 2013.

[138] E. Hosseini, C. Grootaert, W. Verstraete, T. Van de Wiele, "Propionate as a health-promoting microbial metabolite in the human gut," *Nutrition Reviews*, vol. 69, no. 5, pp. 245-258, 2011.

[139] M. Dalmasso, J. Aubert, V. Briard-Bion, V. Chuat, S.-M. Deutsch, S. Even, H.

Falentin, G. Jan, J. Jardin, M.-B. Maillard, S. Parayre, M. Piot, J. Tanskanen, and A. Thierry, "A temporal -omic study of *Propionibacterium freudenreichii* CIRM-BIA1^T adaptation strategies in conditions mimicking cheese ripening in the cold," *PLoS ONE*, vol. 7, no. 1, e29083, 2012.

[140] G. G. Pritchard, J. W. T. Wimpenny, H. A. Morris, M. W. A. Lewis, and D. E. Hughes, "Effects of oxygen on *Propionibacterium shermanii* grown in continuous culture," *Journal of General Microbiology*, vol. 102, no. 2, pp. 223-233, 1977.

[141] G. Zárate, "Dairy propionibacteria: less conventional probiotics to improve the human and animal health," *INTECH Open Access Publisher*, 2012.

[142] S. Lee, I. Kohane, and S. Kasif, "Genes involved in complex adaptive processes tend to have highly conserved upstream regions in mammalian genomes," *BMC Genomics*, vol. 6, no. 168, 2005.

[143] E. Missiaglia, M. Donadelli, M. Palmieri, T. Crnogorac-Jurcevic, A. Scarpa, and N. R. Lemoine, "Growth delay of human pancreatic cancer cells by methylase inhibitor 5-aza-20'-deoxycytidine treatment is associated with activation of the interferon signalling pathway," *Oncogene*, vol. 24, pp. 199-211, 2005.

[144] F. Luft, "Preventing autoimmunity by regulating regulatory T-cell induction," *Journal of Molecular Medicine*, vol. 87, no. 12, pp. 1153-1156, 2009.

[145] E. Kakariari, M. D. Georgalaki, G. Kalantzopoulos, and E. Tsakalidou, "Purification and characterization of an intracellular esterase from *Propionibacterium freudenreichii* ssp. *freudenreichii* ITG 14," *Le Lait*, vol. 80, no. 5, pp. 491-501, 2000.

[146] D. A. Brede, T. Faye, M. P. Stierli, G. Dasen, A. Theiler, I. F. Nes, L. Meile, and H. Holo, "Heterologous production of antimicrobial peptides in *Propionibacterium freudenreichii*," *Applied and Environmental Microbiology*, vol. 71, no. 12, pp. 8077-8084, 2005.

[147] M. T. Fröhlich-Wyder and H. P. Bachmann, "Cheeses with propionic acid fermentation," in *Cheese: Chemistry, Physics and Microbiology*, P. F. Fox, P. L. H. McSweeney, T. M. Cogan, and T. P. Guinee, Eds., pp. 141-156, Elsevier Academic

Press, Oxford, UK, 3rd edition, 2004.

[148] S. J. Lee, H. Song, and S. Y. Lee, "Genome-based metabolic engineering of *Mannheimia succiniciproducens* for succinic acid production," *Applied and Environmental Microbiology*, vol. 72, no. 3, pp. 1939-1948, 2006.

[149] T. Kurzrock and D. Weuster-Botz, "Recovery of succinic acid from fermentation broth," *Biotechnology Letters*, vol. 32, no. 3, pp 331-339, 2010.

[150] R. S. Igoe and Y. H. Hui, *Dictionary of Food Ingredients*, Aspen, MD, USA, 2001.

[151] C. Du, S. K. Lin, A. Koutinas, R. Wang, P. Dorado, and C. Webb, "A wheat biorefining strategy based on solid-state fermentation for fermentative production of succinic acid", *Bioresource Technology*, vol. 99, no. 17, pp. 8310-8315, 2008.

[152] H. Song, Y. S. Huh, S. Y. Lee, W. H. Hong, and Y. K. Hong, "Recovery of succinic acid produced by fermentation of a metabolically engineered *Mannheimia succiniciproducens* strain," *Journal of Biotechnology*, vol. 32, no. 4, pp. 445-452, 2007.

[153] P. C. Lee, W. G. Lee, S. Y. Lee, and H. N. Chang, "Succinic acid production with reduced by-product formation in the fermentation of *Anaerobiospirillum succiniciproducens* using glycerol as a carbon source," *Biotechnology and Bioengineering*, vol. 72, no. 1, pp. 41-48, 2001.

[154] S. Varadarajan and D. J. Miller, "Catalytic upgrading of fermentation-derived organic acids," *Biotechnology Progress*, vol. 15, no. 5, pp. 845-854, 1999.

[155] V. B. Jayaram, S. Cuyvers, K. J. Verstrepen, J. A. Delcour, and C. M. Courtin, "Succinic acid in levels produced by yeast (*Saccharomyces cerevisiae*) during fermentation strongly impacts wheat bread dough properties," *Food Chemistry*, vol. 151, no. 15, pp. 421-428, 2014.

[156] B. Cok, I. Tsiropoulos, A. L. Roes, and M. K. Patel, "Succinic acid production derived from carbohydrates: an energy and greenhouse gas assessment of a platform chemical toward a bio-based economy," *Biofuels, Bioproducts and Biorefining*, vol.

8, no. 1, pp. 16-29, 2014.

[157] B. A. Wilson, J. C. Schisler, and M. S. Willis, "Sir Hans Adolf Krebs: architect of metabolic cycles," *LabMedicine*, vol. 41, no. 6, pp. 377-380, 2010.

[158] J. Vélez, N. H. Jr., M. Konopleva, Z. Zeng, K. Kojima, I. Samudio, and M. Andreeff, "Mitochondrial uncoupling and the reprograming of intermediary metabolism in leukemia cells," *Frontiers in Oncology*, vol. 3, no. 67, 2013.

[159] D. Voet and J. G. Voet, *Biochemistry*, John Wiley & Sons, Inc., Hoboken, NJ, USA, 2011.

[160] T. J. Erb, "Carboxylases in natural and synthetic microbial pathways," *Applied and Environmental Microbiology*, vol. 77, no. 24, pp. 8466-8477, 2011.

[161] A. Zhang, J. Sun, Z. Wang, S.-T. Yang, and H. Zhou, "Effects of carbon dioxide on cell growth and propionic acid production from glycerol and glucose by *Propionibacterium acidipropionici*," *Bioresource Technology*, vol. 175, pp. 374-381, 2015.

[162] Z. Wang, M. Lin, L. Wang, E. M. Ammar, and S.-T. Yanga, "Metabolic engineering of *Propionibacterium freudenreichii* subsp. *shermanii* for enhanced propionic acid fermentation: effects of overexpressing three biotin-dependent carboxylases," *Process Biochemistry*, vol. 50, no. 2, pp. 194-204, 2015.

[163] C. Thakker, I. Martínez, W. Li, K.-Y. San, and G. N. Bennett, "Metabolic engineering of carbon and redox flow in the production of small organic acids," *Journal of Industrial Microbiology and Biotechnology*, vol. 42, no. 3, pp. 403-422, 2015.

[164] D. J. V. Beste, B. Bonde, N. Hawkins, J. L. Ward, M. H. Beale, S. Noack, K. Nöh, N. J. Kruger, R. G. Ratcliffe, and J. McFadden, "^{13}C metabolic flux analysis identifies an unusual route for pyruvate dissimilation in mycobacteria which requires isocitrate lyase and carbon dioxide fixation," *PLoS Pathogens*, vol. 7, no. 7, e1002091, 2011.

[165] Y. Yamauchi, T. Hirasawa, M. Nishii, C. Furusawa, and H. Shimizu,

"Enhanced acetic acid and succinic acid production under microaerobic conditions by *Corynebacterium glutamicum* harboring *Escherichia coli* transhydrogenase gene *pntAB*," *The Journal of General and Applied Microbiology*, vol. 60, no 3, pp. 112-118, 2014.

[166] D. J. Kelly and N. J. Hughes, "The citric acid cycle and fatty acid biosynthesis," in *Helicobacter pylori: Physiology and Genetics*, H. L.T. Mobley, G. L. Mendz, and S. L. Hazell, Eds., pp. 135-146, ASM Press, Washington, D.C., USA, 2001.

[167] L. Vadlakonda, A. Dash, M. Pasupuleti, K. A. Kumar, and P. Reddanna, "Did we get Pasteur, Warburg, and Crabtree on a right note?," *Frontiers in Oncology*, vol. 3, no. 186, 2013.

CHAPTER 4

BIOMEDICAL ASPECTS OF FERMENTED FOODS

4.1 Nutritional concepts and definitions applied to fermented foods

4.1.1 The functional food concept: applicability to fermented foods

The history of probiotics began the first years of the 1900, when Ilya Ilyich Mechnikov (1845-1916), Nobel Prize in Physiology or Medicine (1908), observed that not all microorganisms were harmful to human health and that some intestinal bacteria produced useful compounds against aging [1, 2].

However, the concept of functional food only arose in the 1960s with the scientific demonstration that the Scandinavian dissociated diet (alternate predominance of certain nutrients, e.g., lipids, then carbohydrates) could enhance muscular function, followed by the Japanese FOSHU (Foods for Specified Health Uses) concept in the late 1980s, and a European consensus on functional food in the late 1990s [3-7].

Most of the authors agreed in defining functional food as any food (or food component) that provides health benefits beyond basic nutrition, using many synonyms, *i.e.* designer food, medicinal food, nutraceutical, therapeutic food, superfood, foodiceutical and medifood [8-13].

Hence, fermented foods are fully consistent with the general definition of functional foods because they provide extra health benefits through their content of bioactive substances (polyphenols, mostly flavonoids) and probiotics, additionally fulfilling the nutritional requirements for minerals (e.g. iron), water-soluble vitamins (e.g. vitamin B_{12} and folate) and fat-soluble vitamins (K_2 or menaquinones) [11, 14-22]. Each of these aspects of fermented food is important and deserved separate consideration in this book.

4.1.2 The concepts of nutrient density and energy density applied to fermented foods

The concept of nutrient density, which has been applied to both individual foods and the total diet, represents the nutritive value relative to the total energy content [23]. Hence, nutrient-dense foods are those that contribute more beneficial nutrients than calories to the overall diet [24]. Diverse examples of vitamins and bioactive molecules present in fermented foods have already been given in the previous pages of this book. Nevertheless, two fermented dairy products clearly suit this definition: yogurts and cheeses [24]. In recent years, these dairy products have been found to have a variety of health-promoting effects, particularly yogurts, as discussed below under the subsection of probiotics [25-27].

Energy density of foods may be defined as the energy per unit weight or volume (kcal/100g or megajoules per kilogram) [28]. In terms of energy-rich macronutrients, carbohydrates predominate in yogurts, whereas fats predominate in cheeses (**Table 2**) [29, 30]. However, both yogurts and cheeses are rich in high-quality proteins, *i.e.* those that are easily digested and contain quantities of the essential amino acids corresponding to human requirements [31]. It is appropriate to review some aspects of these two categories of fermented dairy products as follows.

The three major constituents of cheeses are casein, fat and water, all of which contribute to the structure and texture of the products [32]. Cheeses may be classified according to many factors, the moisture (wt%) being one of the most important: very-hard cheese (wt% < 25), e.g. Parmesan; hard cheeses (wt% 25-36), e.g. Emmental, Gruyère; semi-hard cheeses (wt% 36-40), e.g. Edam, Gouda; soft cheeses (wt% > 40), e.g. Camembert, Roquefort [33].

Hard cheeses are produced using an enzyme called chymosin (EC 3.4.23.4), which is found in rennet, which is, in turn, produced from the stomachs of un-weaned calves[++++++] [34, 35]. Hard cheeses possess high dry matter content, usually aged for a number of years and pressed with weights during ripening to extract whey [36]. Whey is the basic liquid by-product remaining after the precipitation and removal of milk casein during cheese manufacturing [37]. Consequently, hard cheeses are not able to absorb water molecules like other products because this property appears typical for high molecular weight casein chains with limited (or absent) proteolysis e.g. in cow milk curds and *pasta filata* cheeses [38]. Hard cheeses are firm in texture, usually waxed or vacuum-packaged and as a rule, the harder the cheese, the longer it keeps [39].

Soft cheeses are creamy, have a smooth texture and are made from milk with a relatively low dry matter content and a range of fat contents (from skim milk soft cheeses having a butterfat content of <2%, to full-fat soft cheeses containing at least 20% butterfat) [36]. Superior quality parameters of soft cheeses are obtained from milk with a dominant proportion of small milk fat globules (MFGs) because they bind water more easily, which results in a desirable softer consistency of the final product (conversely, milk with a dominant share of large MFGs is more suitable for the manufacture of hard cheeses) [40]. It is worth noting that hard cheeses are lower in lactose[§§§§§§] than other cheeses and may be better tolerated than soft cheeses [41, 42]. Soft cheeses are frequently characterized by ripening from the surface to the interior, mainly through the action of microorganisms present in the smear [43].

Ripening is another biochemical process that enriches the sensory composition

---

[++++++]Rennet sources include animals, plants and microorganisms, according to this reference [35].

[§§§§§§]Lactose is one of the most common disaccharides in the human diet and is composed of glucose and galactose linked by α-(1,4)-glycosidic bond, representing the principal sugar in milk (including maternal breast milk, standard infant formula and cow's milk) [36, 41, 42]. While congenital lactase (EC 3.2.1.108) deficiency is rare, developmental lactase nonpersistence (adult-type hypolactasia) resulting from genetic hard wiring occurs such that most people experience a maximal lactase production at birth that declines in childhood or adolescence [41]. The majority of individuals from African, Arab and Hispanic descent and in the near totality of individuals from Asian descent experience lactase nonpersistence, whereas the prevalence of lactase deficiency among Europeans is <5% [41]. Fermented dairy products usually have less lactose because this disaccharide has been partially used by the bacteria to produce the desirable flavors and textures of products like yogurts and cheeses [41, 42].

of cheeses. It is a time-consuming process involving complex and well-balanced reactions between glycolysis, proteolysis and lipolysis of the milk components [39].

The flavor compounds commonly found in smear-ripened cheeses are produced by the cheese microbiota (e.g. *Corynebacterium variabile* DSM 44702) as a diverse group of volatile substances (e.g. sulphur-containing compounds, esters, aldehydes and ketones); although half of these compounds derives from lactose fermentation, citrate degradation and a few from lipolysis and the second half derives from amino acid catabolism by complex metabolic pathways [44].

Changes in cheese color are also complex. For instance, red-smear ripened soft cheeses are characterized by their orange-red color which may originate from pigments produced by yellow bacteria (e.g. *Microbacterium gubbeenense* or *Arthrobacter arilaitensis* or yet *Brevibacterium linens* and related species such as *B. aurantiacum*) and β-carotene derived from milk [45].

Additionally, the bioactive peptides in cheeses exert biological activities (e.g. antihypertensive and antioxidative) through casein segments (αS1 1-13 and αS1 4-13, respectively) [25]. Thus, fermentation enhances the nutrient content of foods not only through the biosynthesis of essential amino acids and proteins, but also through the biosynthesis of vitamins, as further discussed in the next subsection of this review [46, 47]. The high value of fermented foods places them into two larger categories of foods: nutrient-dense and energy-dense foods (**Table 2**).

Table 2:
General nutritional composition of selected cheeses and yogurts (adapted from [29, 30]).

| Cheeses | Components (g/100g fresh weight) | | | | | | Energy | |
	Lipids	Carbohydrates	Proteins	Ash (■)	Moisture (%)	Cholesterol (mg) (●)	kcal	kJ
Parmesan	33.5	1.7	35.6	8.0	21.2	106	453	1895
Emmental	29.7	Traces	28.7	-	35.7	90	382	1587
Gruyère	33.3	Traces	27.2	-	35.0	100	409	1695
Edam	25.4	Traces	26.0	-	43.8	80	333	1382
Gouda	31.0	Traces	24.0	-	41.1	100	375	1555
Camembert (♦)	23.7	Traces	20.9	-	50.7	75	297	1232
Roquefort (♦)	32.9	Traces	19.7	-	41.3	90	375	1552
Yogurts								
Natural	3.0	1.9	4.1	0.9	90.0	14	51	215
Low fat	0.3	5.8	3.8	0.9	89.2	3	41	174
Strawberry flavor	2.3	9.7	2.7	0.6	84.6	7	70	291
Peach flavor	2.3	9.4	2.5	0.6	85.1	8	68	284

(♦): due to the low mechanical resistance, which further decreases during ripening, soft cheeses are usually covered in secondary packages made from cardboard or wood [48].
(■): ash content represents the total mineral amount in foods [49].
(-): not determined.
(●): the recommended daily intake of cholesterol for patients at risk of vascular disease is less than 200 mg/day [50].

4.2 Nutritional value and health benefits of fermented foods

4.2.1 Minerals

Minerals may be defined as natural components formed through geological processes needed in small amounts to regulate body functions [51]. According to the body needs, minerals may be divided into macrominerals (major elements) or microminerals (trace elements) [52]. Macrominerals may be defined as minerals that are required in amounts greater than 100 mg/dL, including sodium (Na), potassium (K), chloride (Cl), calcium (Ca), magnesium (Mg), phosphorus (P) [19, 52]. Microminerals are those inorganic constituents of the body which are required at less than 100 mg/dL (in other word, it forms less than 0.01% of the body weight), including copper (Cu), iron (Fe), zinc (Zn), selenium (Se), molybdenum (Mo), fluorine (F), iodine (I), manganese (Mn), cobalt (Co) – **Table 3** [19, 52].

Several factors affect the content of minerals in fermented foods, e.g. the raw materials used for fermentation (e.g. plants, animals, types of milk), the fermentation processes and the microorganisms applied in such processes (e.g. lactic acid bacteria), the subsequent physicochemical changes (e.g. pH changes), among many others.

Table 3:
Mineral composition of selected cheeses and yogurts (adapted from [29, 30]).

Cheeses	Macrominerals (mg/100 g)					Microminerals (mg/100 g)			
	Na	K	Ca	Mg	P	Fe	Zn	Cu	Mn
Parmesan	1090	110	1200	45	810	1.1	5.3	0.17	0.05
Emmental	450	89	970	35	590	0.3	4.4	-	-
Gruyère	670	99	950	37	610	0.3	2.3	-	-
Edam	1020	97	770	39	530	0.4	2.2	-	-
Gouda	910	91	740	38	490	0.1	1.8	-	-
Camembert	650	100	350	21	310	0.2	2.7	-	-
Roquefort	1670	91	530	33	400	0.4	1.6	-	-
Yogurts									
Natural	52	71	143	11	119	Traces	0.4	0.02	Traces
Low fat	60	182	157	12	110	Traces	0.5	Traces	Traces
Strawberry flavor	38	52	101	8	73	Traces	0.3	0.02	Traces
Peach flavor	37	52	95	8	66	0.1	0.3	0.02	Traces

(-): not determined.

Iron is a vital mineral and there are two main types of iron available in foods: the heme iron, present in meat and its products (around 20-30% being absorbed) and the non-heme iron, existent in plants, with a much lower bioavailability (1-8%) [53]. Calcium, magnesium, sodium, potassium, iodine and selenium do not interfere with iron absorption [53]. Divalent metal transporter 1 (or DMT1), the main iron transporter, may be involved in copper transport in low iron conditions [53-55]. Manganese inhibits iron absorption [53]. Phytate, soy protein, calcium and polyphenols inhibit non-heme iron absorption, whereas ascorbate, meat/fish and fermented foods enhance it [53].

It was recently demonstrated that the ferric iron (Fe^{3+}) to ferrous iron (Fe^{2+}) ratio is changed after lactic acid fermentation of vegetables [55]. Although Fe^{2+} is the transported species, it may be favorable for iron absorption to have Fe^{3+} in the gastrointestinal passage since this species is less reactive than its reduced counterpart; additionally, the mucus layer binds Fe^{3+} under acidic conditions, which may provide the reductase (in the enterocyte membrane) with its substrate; essential for iron absorption [52, 55]. To close, iron, together with dietary fibers (particularly phytate), calcium and pharmacological intakes of folic acid can interfere with zinc absorption [53].

Zinc is an essential micromineral involved in a wide array of metabolic actions such as acid-base balance, amino acid metabolism, protein synthesis, zinc-finger proteins, nucleic acid synthesis, folate availability, vision, the immune system, reproduction, the activation of many enzymes, somatic growth and the development of the nervous system [56, 57]. Copper is another essential micromineral that functions in a diverse array of biochemical processes that include mitochondrial respiration, neurotransmitter biogenesis, connective tissue maturation and reactive oxygen chemistry [58]. Zinc and copper are generally found in the same foods (liver and seafood such as oysters) [59]. The recommended dietary intake of zinc ranges between 8-15 mg/day [19]. The recommended dietary intake of copper for adult men and women is 900 μg/day [60].

Cassava is an important starchy substrate used for preparing fermented foods in the world, especially in African countries, e.g. *Kpukpuru* (a fermented cassava staple) in Nigeria, *attieke* (another cassava-based food product) in Benin [61, 62]. Nevertheless, cassava contains little zinc and children who consume it as a staple food are at risk for inadequate intake of not only zinc but also of iron and vitamin A [63]. Chronic zinc deficiency may result in reduced growth (dwarfism), inhibited sexual development, and it must be treated with adequate zinc intake [64].

The consumption of fermented foods leads to enhanced zinc absorption most likely due to the presence of organic acids (acetic, citric, lactic or malic acids) that lower the pH and provide the optimal conditions for the enzymatic degradation of the anti-nutrient phytate (a mineral chelator), thereby increasing the solubility and bioavailability of zinc through this pro-enzymatic action [56, 59, 65-68].

Furthermore, the uptake of zinc by the probiotic lactic acid bacteria (e.g. *Lactobacillus acidophilus* WC 0203), could function as organic matrices for zinc incorporation and provide a future basis for optimization of zinc supplementation or fortification [69, 70].

The fermentation of sugared tea with a symbiotic culture of acetic acid bacteria and yeast (tea fungus) yields kombucha tea, which is consumed worldwide for its refreshing and beneficial properties on human health (the determination of minerals such as copper, iron, manganese, nickel and zinc, showed their increase probably as a result of the metabolic activity of kombucha[*******]) [71].

As mentioned above, the food processing may affect and be more critical to the availability of the minerals. For instance, the effect of fermentation on in vitro mineral estimation of selected Indian foods increased the bioavailability of zinc, iron, calcium and magnesium [72]. In contrast, the availability of zinc in the autoclaved black-gram (*Phaseolus mungo* L., Fabaceae) diet (in rats) was better as compared to the germinated, fermented and raw black-gram diets and this may be due to more destruction of phytate [73].

Manifestations of zinc toxicity consist of an impaired immune response, the reduction of high-density lipoprotein (HDL) levels and the induction of copper deficiency anemia, which is indistinguishable from iron-deficiency anemia [59, 64, 74]. The Tolerable Upper Intake Levels (ULs) of zinc for children (aged 4-8 years) and adults (aged $\geq$ 19 years) are 12 mg/day and 40 mg/day, respectively [75]. The toxic effects of high doses of zinc are uncommon, although there is conflicting information in the literature according to which dietary intakes of 18.5-25 mg/day (inferior to ULs for adults) have been reported to impair the copper status of adult subjects [19, 59, 75]. Other clinical manifestations of copper deficiency include neutropenia, hypochromic microcytic anemia, depigmentation of skin and hair,

---

[*******]It originated in northeast China (Manchuria) where it was prized during the Tsin Dynasty ("Ling Chi"), about 220 BC, for its detoxifying and energizing properties [71].

neurologic disturbances, lethargy and abnormalities of connective tissue accompanied by skeletal abnormalities [59]. In contrast, copper toxicity (Wilson's disease) has an excellent prognosis if diagnosed early and treated properly [76].

4.2.2 Vitamin B_{12} and folate

Vitamin B_{12} or cabalamin (**Figure 19**) and folate or folic acid or yet vitamin B_9 (**Figure 20**) are water-soluble vitamins essential for proper metabolic functioning in animals, including humans [77, 78].

Plants are devoid of vitamin B_{12}, whereas folate is particularly plentiful in fresh green vegetables, liver and some fresh fruits [59, 79].

Among the B_{12}-producing species are the following genera: *Aerobacter, Agrobacterium, Alcaligenes, Azotobacter, Bacillus, Bifidobacterium, Citrobacter, Clostridium, Corynebacterium, Flavobacterium, Hodgkinia, Klebsiella, Micromonospora, Mycobacterium, Norcardia, Propionibacterium, Protaminobacter, Proteus, Pseudomonas, Rhizobium, Salmonella, Serratia, Streptomyces, Streptococcus, Thermus,* and *Xanthomonas* [80-84]. Bacteria such as *Lactobacillus reuteri* JCM1112 advantageously produce both vitamin B_{12} and folate in fruit fermentation [18].

From an industrial perspective, vitamin B_{12} is obtained exclusively by the fermentation process and is produced by a number of pharmaceutical companies to meet the annual demands worldwide [85].

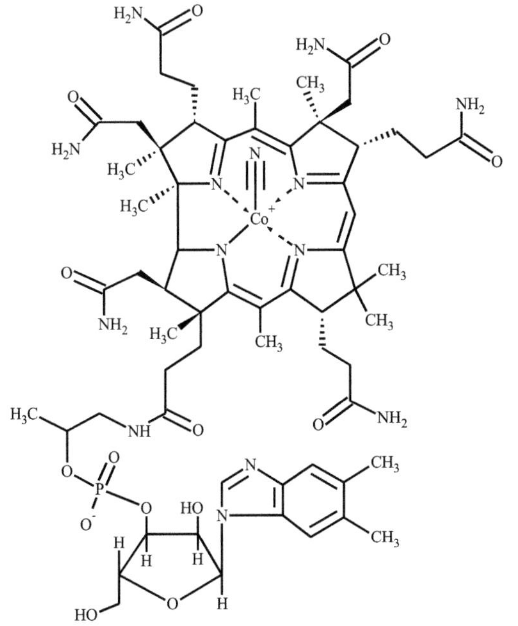

Figure 19:
Vitamin B$_{12}$ (or cabalamin).
It is a cobalt-containing modified tetrapyrrole that is an essential nutrient for higher animals; its biosynthesis is restricted to certain bacteria and requires approximately 30 enzymatic steps for its complete *de novo* construction [77]. Two distinct biosynthetic pathways exist, which are termed the aerobic and anaerobic routes, both fully detailed now [77, 86].

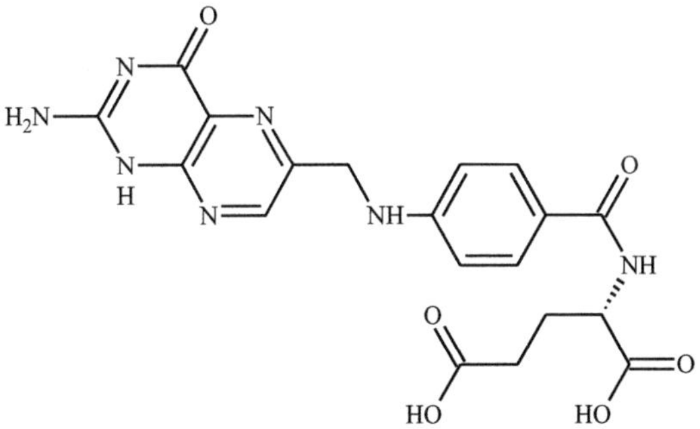

Figure 20:
Folic acid (N-(4-{[(2-amino-4-oxo-1,4-dihydropteridin-6-yl)methyl]amino}benzoyl)-L-glutamic acid). The occurrence of folic acid in nature is not in appreciable amounts, though it is assimilated in the body and is converted to the active cofactor form of the vitamin, tetrahydrofolate [87-89].

Therefore, humans have an auxotrophic requirement for vitamin B_{12} and folate, and the recommended intakes of these nutrients for healthy adults are 2.4 and 400 µg/day, respectively [19, 90]. Nonetheless, individuals cannot absorb vitamin B_{12} if sufficient intrinsic factor is not available in the stomach [19]. Consequently, one of the most common causes of vitamin B_{12} deficiency is the inadequate secretion of intrinsic factor rather than the inadequate dietary intake of the vitamin only [19, 90]. The primary cause of folate deficiency is the low intake of sources rich in the vitamin such as legumes and green leafy vegetables and the consumption of these foods may explain why folate status can be adequate in relatively poor populations (other situations in which the risk of folate deficiency increases, include lactation and alcoholism) [90]. Folate deficiency results in megaloblastic anemia that cannot be distinguished from that caused by a deficiency of vitamin B_{12} [59].

There are various fermented foods around the world that could augment vitamin B_{12} and folate in human diets. For instance, *kimchi* is a traditional Korean food that is fermented from vegetables such as Chinese cabbage and radish (*Raphanus sativas* L., Brassicaceae), in which lactic acid bacteria are known to perform significant roles [91, 92]. Nevertheless, *Leuconostoc mesenteroides* is the dominant microorganism before ripening and *Lactobacillus* species may be dominant in the later stages of *kimchi* fermentation depending on the temperature [92]. Importantly, the vitamin B_{12} content (µg/100g dry weight; Mean ± SD) in *kimchi* varies according to the plant and to its ripening stage: cabbage *kimchi* (0.24 ± 0.01), mustard[†††††††] leaves *kimchi* (0.13 ± 0.02), yeolmu *kimchi* or young radish *kimchi* (0.12 ± 0.09), radish *kimchi* (0.04 ± 0.01) [93-95]. *Kimchi* and rice are the main sources of folate in Koreans, which contributed about 30-40% of folate intake and could explain the low prevalence of serum folate deficiency in this country [96].

Injera is an Amharic word for an Ethiopian bread made from teff (*Eragrostis*

[†††††††]Mustard belongs to the botanical family of Cruciferae or Brassicaceae, comprising two major types of seeds, the white or yellow (*Sinapis alba* L., Brassicaceae), as a main ingredient of the American-style mustards, and the brown or oriental mustard (*S. juncea* L., Brassicaceae) that is used for European and Chinese products [95].

tef (Zuccagni) Trotter, Poaceae), wheat, barley, maize, sorghum or from a mixture of them [97]. From the nutritional point of view, teff is a highly valuable grain as it contains high levels of calcium, magnesium and iron as well as folate [98]. Injera is similar in shape and texture to the pancake of the United States [97]. However, a study of vitamin composition of ethnic foods commonly consumed in Israel demonstrated that injera possessed folate and vitamin B_{12} ($\mu g/100$ g of edible portion) below detectable levels [99]. These levels could be partially explained by the baking process, which would be able to reduce the amount of (heat labile) folate isoforms [100, 101]. Moreover, there are contradictory related data in the literature, describing folate and vitamin B_{12} as heat stable molecules [102]. Lastly, *tempeh* is a fermented soyfood from Indonesia that has a much higher content in vitamin B_{12} and folate than unfermented soybeans and soybean-fermented foods contribute significantly to the good vitamin B_{12} status of centenarian and very old Koreans (aged 85 or more) [16, 17, 103].

Thus, certain fermented foods are rich in vitamin B_{12} and folate and should be recommended along with other food sources (meat products), particularly in the case of deficiency of these two vitamins [59].

4.2.3 Vitamin K

The term vitamin K denotes a group of lipophilic, hydrophobic vitamins that belong to the class of 2-methyl-1,4-naphthoquinone derivatives [104]. All of the members of the vitamin K group share a common methylated naphthoquinone ring structure, but have different aliphatic side chains attached at the 3-position [104]. This vitamin group includes K_1 (or phylloquinone), K_2 (or menaquinones) and K_3 (or menadione) [19, 57, 105]. Green vegetables are the main dietary sources of phylloquinone (**Figure 21**), contributing up to 60% of the total phylloquinone intake [105, 106].

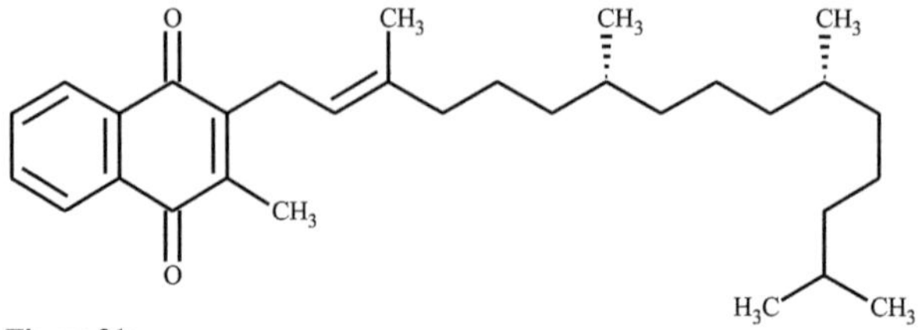

Figure 21:
Vitamin K_1 (or phylloquinone).
A recent study demonstrated that phylloquinone supplementation for four weeks improved glycemic status in premenopausal prediabetic women independent of adiponectin (as further discussed in this subsection, osteocalcin increases ß-cell proliferation as well as insulin and adiponectin secretion, which improve glucose tolerance and insulin sensitivity) [106].

Vitamin K_2 comprises a family of compounds, generically called menaquinones (MKs), the side chain of which consists of repeated (5-carbon) isoprene units, from 1 to 14 of them, and depending on the number of isoprene units present, they are referred to as MK1-MK14 [107, 108]. In the United States, poultry products are the primary dietary sources of MK-4 because poultry feed is a rich source of menadione, which is subsequently converted to MK-4 in certain tissues, whereas in Japan, a unique source of MK-7 is the traditional food *natto*, a fermented soybean product [105, 109, 110]. *Bacillus subtilis natto* is the source of vitamin K_2 (almost exclusively MK-7) in this food, which may contain up to 1100 mg of K_2 per 100 grams of food [109]. Other vitamin K_2-producing bacteria are lactic acid bacteria (mainly MK-8 and MK-9) and propionic acid bacteria (mainly MK-10) in the production of cheese and curd cheese in Europe and northern America [109]. Interestingly, the sharp taste of *natto* is highly appreciated in Japan, but not in western societies [109]. For decades, health benefits have been associated with the consumption of *natto*, including the stimulation of the immune system [110]. It is also known that soy foods, such as soy sauce, *miso*[‡‡‡‡‡‡‡] and *natto*, show an antihypertensive effect by inhibiting the angiotensin-converting enzyme (ACE), the key enzyme in the renin-angiotensin system (**Figure 22**) [111, 112].

[‡‡‡‡‡‡‡]*Miso* is generally prepared with *koji* (barley grains fermented by fungi belonging to the *Aspergillus* genus), steamed beans, water, salt and a sufficient quantity of starters (halotolerant yeasts and lactic acid bacteria) and finally skimmed milk (a product of Holstein cows) [112].

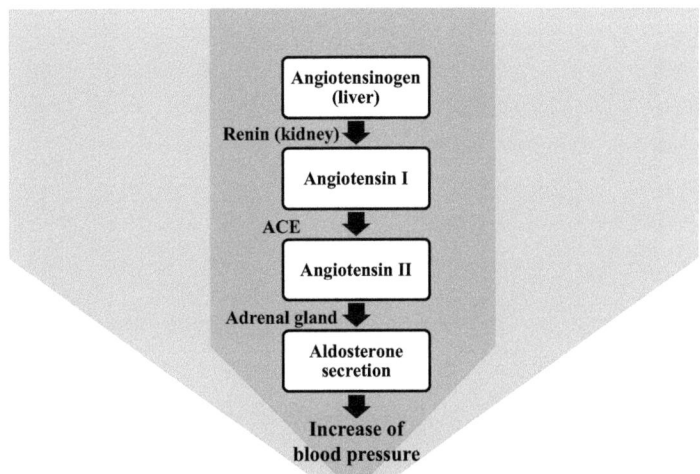

Figure 22:
Simplified scheme of the renin–angiotensin system (adapted from [111]).

According to Cutting [110], "nattokinase"[§§§§§§] is a serine protease produced by all strains of *B. subtilis*, but in the *natto* strain, it is produced at high levels; moreover, it cannot be ruled out that health benefits ascribed to *natto* require consumption of both soybeans and bacteria, rather than just the bacterium. It is important to know the vitamin K content of common fermented and non-fermented foods (**Table 4**).

[§§§§§§]Nattokinase has GRAS (Generally Regarded as Safe) status as an enzyme produced from a bacterium in the USA and is purified and sold as a health supplement worldwide [110].

Table 4:
Vitamin K content of common foods (adapted from [113, 114]).

Vegetables	Vitamin K forms	Concentration (μg/100 g)
Collards (*Brassica oleracea* L., Brassicaceae)	Phylloquinone	440
Spinach (*Spinacia oleracea* L., Amaranthaceae)	Phylloquinone	380
Broccoli ********	Phylloquinone	180
Cabbage	Phylloquinone	145
Fermented foods		
Natto	Menaquinone-7	998
Hard cheeses	Menaquinone-9	51.1
Soft cheeses	Menaquinone-9	39.5

MKs are also synthesized by several bacterial species, such as the mutualistic strains of *Escherichia coli* that naturally occur in the large intestine of animals, including humans [19, 57, 115-118].

Menadione is a synthetic form of the vitamin that is clinically utilized as a coagulant [19].

Vitamin K compounds are essential for the function of several proteins involved in blood coagulation (prothrombin, also known as factor II, factors VII, IX and X, protein C, protein S and protein Z), bone metabolism (as discussed below), as well as vascular biology, cell growth and apoptosis (growth-arrest-specific gene 6 protein) [104].

Bone tissue is primarily composed of proteins and minerals (80-90% calcium and phosphorus) and the two most abundant proteins in bone tissue are collagen and osteocalcin [119, 120]. Osteocalcin is a 5.8 kDa hydroxyapatite-binding protein that is exclusively synthesized by osteoblasts, odontoblasts and hypertrophic chondrocytes, which is considered as a specific marker of osteoblast (bone-forming cell) function [121, 122]. One of the most important characteristics of osteocalcin is its three vitamin K-dependent, γ-carboxyglutamic acid residues, which are responsible for its calcium binding properties [122]. This basic finding has generated

---

********Broccoli belongs to the Brassicacea family and is considered among other cole vegetable crops, such as cabbage, cauliflower, Chinese cabbage, Brussels sprouts and kohlrabi [114].

much interest in the role of vitamin K in bone health and disease and has even paved the way to discover the role of vitamin K in the endocrine system and energy metabolism (**Figure 23**) [123].

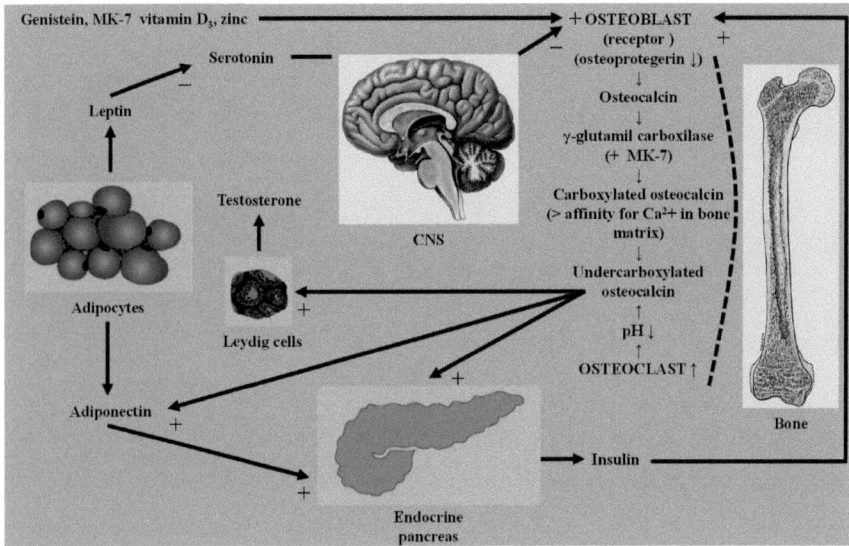

Figure 23:
Endocrine and metabolic roles of osteocalcin (adapted from [124-128]).
Arrows followed by positive sign (+) indicate stimulation. Arrows followed by negative sign (-) indicate inhibition. The dotted arrow indicates that insulin signaling in osteoblast causes a decrease (↓) in osteoprotegerin expression and enhances (↑) bone resorption (an indirect action on osteoclast) and osteocalcin decarboxylation [124, 125]. The undercarboxylated osteocalcin increases insulin secretion in β-cells, increases insulin sensitivity in muscle, liver and adipose tissue and increases testosterone production in Leydig cells [124, 125]. CNS: central nervous system. Leptin inhibits bone formation through a serotonin central relay (brainstem to hypothalamus) and through hypothalamus generated sympathetic tone [126]. The osteogenic effects of vitamin D_3, MK-7 and genistein have been found to be synergistically enhanced by the combination of zinc, in addition to potential suppressive effects on osteoclastic bone resorption [128].

Osteoporosis is a major bone disease currently defined as a systemic skeletal disease that is characterized by a loss of bone mass and the deterioration of bone microarchitecture, leading to increased susceptibility to fracture; which is a significant public health issue, particularly as the population ages [105, 119, 129-132]. In the light of the discussion above, it is clear today that fermented foods must integrate other alimentary sources rich in vitamin K. For women of post-menopausal age, 180-350 μg/day of vitamin K2-7 may need to be supplemented along with the recommended intake of calcium, magnesium, vitamin D and a balanced diet [133, 134]. In addition, the research about the cardiovascular protection by vitamin K is an evolving field in which a boost of novel and relevant evidence is expected [135].

4.2.4 Polyphenols

Polyphenols are plant secondary metabolites generally involved in their adaptation to environmental stress conditions (e.g. ultraviolet radiation and pathogenic attack) that have evolutionary importance [136-138]. They are organic compounds that have one aromatic nucleus with at least two phenolic functions, particularly in the ortho or para positions [139].

As a result, they are classified on the basis of the number of phenolic rings that they contain, for example, ortho-diphenols (caffeic and chlorogenic acids), triphenols (gallic acid and its esters) and so on [140-142]. They are functionally classified as antioxidants [143]. As such, polyphenols are potently capable of delaying, inhibiting or preventing the oxidation of oxidizable materials by scavenging free radicals and diminishing oxidative stress [144].

Flavonoids constitute the most important single group of polyphenols and their main subclasses are represented by the anthocyanidins (e.g. delphinidin and pelargonidin), flavanols (e.g. catechin and epicatechin), flavanones (e.g. naringenin and hesperetin), flavanonols (e.g. astilbin and engeletin), flavonols (e.g. kaempferol and quercetin), flavones (e.g. apigenin and luteolin) and isoflavones (e.g. daidzein and genistein) [145-149].

Polyphenols, which are found in high concentrations in wines (e.g. resveratrol), teas, grapes and a wide variety of other plants, have been associated with the prevention of heart disease and cancer (**Figure 24**) [150, 151].

Figure 24:
The French paradox. This phenomenon is thought to be explained by a high red wine consumption by the French; red wine being rich in various polyphenolic compounds [151].

Anthocyanidins are predominantly found in red, purple and blue berries and play a pivotal role in the chemopreventive effects attributed to fruits and vegetables [145, 152]. Flavanols, which are found mostly in teas, red grapes, red wines and cocoa-derived products, have been demonstrated to be effective in improving endothelial function and decreasing blood pressure in humans and animals [145, 153]. However, catechins are very reactive when exposed to atmospheric oxygen, are spontaneously oxidized by oxygen in aqueous solutions and are dramatically reduced, by 85%, during black tea manufacturing [154-156]. **Figure 25**.

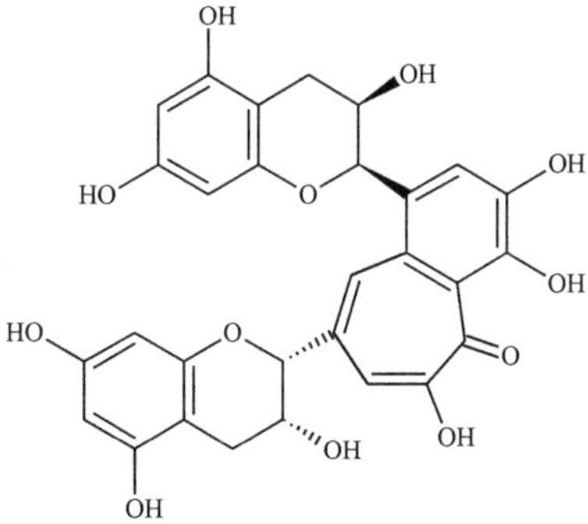

Figure 25:
Theaflavin. It is the plain black tea quality parameter responsible for the astringency (briskness), brightness and color of black tea [157, 158]. Theaflavin is the product of oxidation compounds under catalytic catechin polyphenol oxidase enzyme [158]. Theaflavin is effective in experimental rheumatoid arthritis [159]. Theaflavin is also able to inhibit the influenza virus replication via interference with the haemagglutinin synthesis [160].

105

Flavanones, also known as citrus flavonoids, which are unsurprisingly found in citrus fruits, are effective against the neurogenic inflammation induced by the organic solvent xylene [121, 145, 161, 162]. Like flavanones, flavanonols are found in citrus fruits and like flavanols, flavanonols are inhibitors of lipoxygenase (EC 1.13.11.12), thereby exerting anti-inflammatory activity [163-165].

Flavonols are nearly ubiquitous in foods [145]. However, quercitrin and rutin are bacterially hydrolyzed in the human gut to quercetin, whereas robinin is converted to kaempferol [166]. The consumption of kaempferol-rich vegetables such as broccoli and spinach, is associated with a reduced risk of ovarian cancer [167]. Interestingly, a high content of quercetin was characterized in a local Brazilian dish known as *maniçoba* (Tupi origin) that is prepared from cassava fresh leaves [168, 169]. Quercetin is a potent antioxidant in vitro, which has been suggested as the possible mechanism by which it provides protection against the oxidative damage to low-density lipoprotein (implicated in atherogenesis) [150, 170, 171]. Cocoa-derived flavonols appear to act as mild blockers of the cystic fibrosis transmembrane conductance regulator [172].

Flavones are found in green leafy spices such as parsley (*Petroselinum crispum* (Mill.) Fuss, Apiaceae), thyme (*Thymus vulgaris* L., Lamiaceae) and celery (*Apium graveolens* L., Apiaceae) and act not only as natural free radical scavengers, but also as anti-inflammatory agents due to interfering with the arachidonic acid pathway [145, 173, 174].

Finally, yet importantly, isoflavones are found in soybeans, soy foods and legumes [145]. These diphenolic compounds bind to estrogen receptors and exert estrogen-like effects under certain experimental conditions, and for this reason, they are classified as phytoestrogens [175]. Notably, the consumption of genistein (**Figure 26**) in the diet has been linked to decreased rates of metastatic cancer in a number of population-based studies [176]. DNA methylation is the major epigenetic phenomenon of eukaryotic genomes that involves the addition of a methyl group to the carbon 5 position of the cytosine ring within the dinucleotide 5'-deoxycytidine-deoxyguanidine [177, 178]. Genistein from soybean (among other sources of polyphenols) might inhibit the development of cancer by the reducing DNA (hyper-) methylation status of critical genes associated with cancer, such as p16 or retinoic acid receptor beta [179].

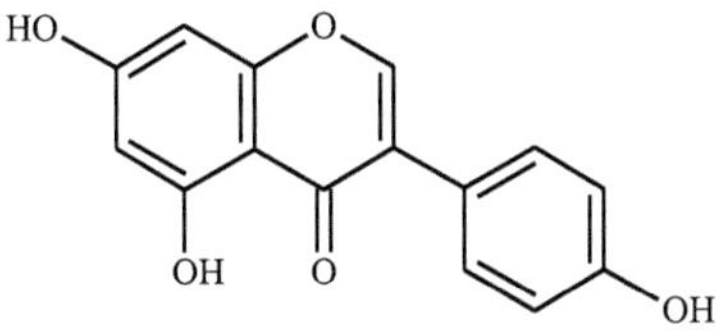

Figure 26:
Genistein (5,7-dihydroxy-3-(4-hydroxyphenyl)-4H-1-benzopyran-4-one): besides its potential benefits abovementioned, it significantly inhibits glycosaminoglycans (GAG) synthesis and decreases GAG storage in fibroblasts of patients suffering from different mucopolysaccharidoses and mucolipidosis types [180].

Thus, according to the discussion above, polyphenols exert pleiotropic (multiple) effects that are primarily related to their molecular and functional variety. They have the advantage of being present in several foodstuffs, particularly worldwide appreciated fermented foods (e.g. red wines, black teas, chocolates and soybean fermented products such as *doenjang*) [181, 182]. Health advances are certainly forthcoming from the multidisciplinary investigations of these compounds.

4.2.5 Probiotics (prebiotics and synbiotics)

The word probiotic is Greek and means "for life" [183]. This term was first used by Lilly and Stillwell in 1965 to describe the substances secreted by one microorganism that stimulates the growth of another [12]. Probiotics are currently defined as living microorganisms present in foodstuffs that exert health effects beyond inherent basic nutrition when ingested in adequate amounts [184-191].

Probiotics are present not only in dairy products, but also in cereals, fruits, vegetables, legumes and meat [192, 193]. With respect to fermented foods, cheeses are effective as functional foods and as convenient vehicles for the introduction of probiotic cultures into the diet because, in comparison with yoghurts and other fermented milk products, cheeses have solid matrices and a higher pH, buffering capacity and fat content, which help protect the probiotic strains during intestinal transit to the sites of action [27, 185, 193-195].

The main probiotic species of the genus *Lactobacillus* are *L. acidophilus, L. bulgaricus, L. casei, L. plantarum, L. reuteri, L. rhamnosus* and *L. paracasei*, whereas those of the genus *Bifidobacterium* are *B. bifidum, B. breve, B. infantis, B. lactis, B. longum* and *B. adolescentis*; other probiotic microorganisms are *Saccharomyces boulardii, Streptococcus thermophilus, Propionibacterium freudenreichii*, and microorganisms of the genera *Bacillus, Escherichia, Enterococcus* and *Pediococcus* [12, 27, 196].

It is worth emphasizing that fermented dairy products are considered the best carriers for probiotic delivery, but other food matrices can maintain probiotic functionality such as vegetable and cereal beverages, among others, as abovementioned [197]. In addition, *S. boulardii* is largely prescribed in Brazil as a commercial lyophilized preparation.

Prebiotics are defined as non-digestible food ingredients (e.g. highly fermentable carbohydrates such as pectins) that beneficially affect the health of the host by selectively stimulating the growth and activity of specific species of bacteria in the gut, particularly the lactic acid-producing bacteria [190, 198-200]. Besides the fermentation of the plant wall structural polysaccharides, a study demonstrated the beneficial impact of glycated pea proteins on the intestinal bacteria from a healthy human, affecting the growth of gut commensal bacteria, particularly lactobacilli and bifidobacteria, whose levels increased significantly [201].

The health benefits are potentially greater when both probiotics and prebiotics are consumed simultaneously (synbiotics), as is the case with the use of probiotic fermented milk containing dietary fiber, which has additive beneficial effects on irritable bowel syndrome and constipation [190, 202]. Probiotics and prebiotics have also been suggested to reduce cholesterol via various mechanisms; although more clinical evidence is needed to strengthen these proposals [200]. Other probiotic benefits include lower frequency and duration of diarrhea associated with antibiotic use and improvement of lactose intolerance [69]. The **Figure 27** summarizes the main effects of the short chain fatty acid fermentation in the gut, emphasizing those of the butyric acid.

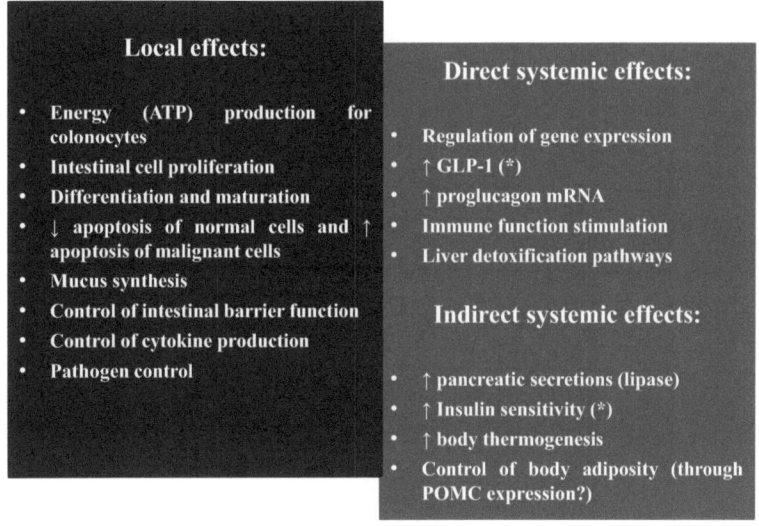

Figure 27:
Effects of butyric acid potentially beneficial to human health (adapted from [203-205]).
↑: increased. ↓: decreased. GLP-1: glucagonlike peptide 1 is an endogenous incretin secreted from the small intestine and capable of activating insulin secretion in healthy individuals. (*): effect mediated by GLP-1 action. POMC: proopiomelanocortine.

Despite the attractive biochemical basis of the short chain fatty acid fermentation in the gut and the resultant (potential) benefits to human health, important pathophysiological questions have been raised regarding their role in non-alcoholic fatty liver disease (NAFLD). This disease is defined as the accumulation of intracellular triacylglycerol in the liver of individuals who consume less than 20 g of alcohol/day [206]. The increased abundance of alcohol-producing bacteria in NAFLD microbiomes, the elevated blood-ethanol concentration in NAFLD patients and the well-established role of alcohol metabolism in oxidative stress and, consequently, liver inflammation, suggest a role for alcohol-producing microbiota in the pathogenesis of NAFLD [207]. Subsequent pertinent data must be demonstrated by appropriate criteria in other works.

4.3 Potential detrimental effects of fermented foods on human health

Biogenic amines are organic bases found in high-protein foods such as fish (e.g. scombroid), meat, fermented products (e.g. cheese, wine, beer) and spoiled foods as a result of a microbial contamination that exhibits amino acid decarboxylase activities [208]. Fermented foods can be considered a reservoir for accumulation of biogenic amines because the fermentation process gives the predominant microbial flora an opportunity to decarboxylate the increasing amount of free amino acids [209]. High levels of biogenic amines in food constitute a potential public concern due to their physiological and toxicological effects [208]. For instance, tryptamine and tyramine have toxic effects in human beings such as blood pressure increase causing hypertension, whereas putrescine and cadaverine are known to enhance histamine toxicity by inhibiting histamine metabolising enzymes (monoamine or diamine oxidase and histamine methyl transferase) [208].

Milk and soft cheeses are highly perishable foods that may be incriminated in *Listeria monocytogenes* infections [210]. Listeriosis can cause severe disease, especially in fetuses, neonates, older adults and persons with certain immunocompromising and chronic conditions [211]. Prevention is needed, including food safety education and messaging (e.g. avoiding Mexican-style cheese during pregnancy) [211].

Botulism cases due to traditional Alaska Native fermented foods occur periodically in Southwest Alaska [212]. Botulism is a rare, but serious illness, caused by a bacterium called *Clostridium botulinum*, which occurs in soil and produces a neurotropic toxin [213]. Foodborne botulism comes from eating foods contaminated with the toxin [213]. Treatment may include antitoxins, intensive medical care or surgery of infected wounds, but timely diagnosis can be lifesaving [213]. Education, facilities for the hygienic preparation of fermented foods and frequent inspection services may help to prevent and control foodborne disease outbreaks [214, 215].

REFERENCES

[1]   G. Caramia and S. Silvi, "Probiotics: from the ancient wisdom to the actual therapeutical and nutraceutical perspective," in *Probiotic Bacteria and Enteric Infections: Cytoprotection by Probiotic Bacteria*, J. J. Malago, J. F. J. G. Koninkx, and R. Marinsek-Logar, Eds., pp. 3-37, Springer, Dordrecht, Netherlands, 2011.

[2]   "Ilya Mechnikov – Facts". Nobelprize.org. Nobel Media AB 2014. Web. 3 Apr 2015.   http://www.nobelprize.org/nobel_prizes/medicine/laureates/1908/mechnikov-facts.html

[3]   J.-M. Antoine, "Fonctionnalité des aliments: un concept à redéfinir?," *Cahiers de Nutrition et de Diététique*, vol. 44, no. 3, pp. 113-116, 2009.

[4]   R. Bailey, "Functional foods in Japan: FOSHU ("Foods for Specified Health Uses") and "Foods with Nutrient Function Claims," in *Regulation of Functional Foods and Nutraceuticals: A Global Perspective*, C. M. Hasler, Ed., pp. 247-261, Blackwell Publishing, IA, USA, 2005.

[5]   F. Brouns, *Essentials of Sports Nutrition*, John Wiley & Sons Ltd, London, England, UK, 2002.

[6]   M. B. Roberfroid, "Concepts and strategy of functional food science: the European perspective," *The American Journal of Clinical Nutrition*, vol. 71, no. 6, pp. 1660s-1664s, 2000.

[7] M. B. Roberfroid, "Global view on functional foods: European perspectives," *British Journal of Nutrition*, vol. 88, supplement S2, pp. S133-S138, 2002.

[8] C. I. Beristain, F. Cruz-Sosa, C. Lobato-Calleros, R. Pedroza-Islas, M. E. Rodríguez-Huezo, and J. R. Verde-Calvo, "Applications of soluble dietary fibers in beverages," *Revista Mexicana de Ingeniería Química*, vol. 5, no. 1, pp. 81-95, 2006.

[9] E. R. Farnworth, "Kefir – a complex probiotic," *The Food Science and Technology Bulletin: Functional Foods*, vol. 2, no. 1, pp. 1-17, 2005.

[10] P. J. Jones, "Clinical nutrition: 7. Functional foods — more than just nutrition," *Canadian Medical Association Journal*, vol. 166, no. 12, pp. 1555-1563, 2002.

[11] B. M. Marriott, "Functional foods: an ecologic perspective," *The American Journal of Clinical Nutrition*, vol. 71, no. 6, pp. 1728s-1734s, 2000.

[12] C. R. Soccol, L. P. S. Vandenberghe, M. R. Spier, A. B. P. Medeiros, C. T. Yamaguishi, J. D. D. Lindner, A. Pandey, and V. Thomaz-Soccol, "The potential of probiotics: a review," *Food Technology and Biotechnology*, vol. 48, no. 4, pp. 413-434, 2010.

[13] M. C. de Andrade Jr., "Berinjela: que antecedente familial terrível!," *Arquivos Brasileiros de Endocrinologia & Metabologia*, vol. 48, no. 4, pp. 572-574, 2004.

[14] A. S. Hummel, C. Hertel, W. H. Holzapfel, C. M. A. P. Franz, "Antibiotic resistances of starter and probiotic strains of lactic acid bacteria," *Applied and Environmental Microbiology*, vol. 73, no. 3, pp. 730-739, 2007.

[15] M. H. J. Knapen, L. J. Schurgers, and C. Vermeer, "Vitamin K_2 supplementation improves hip bone geometry and bone strength indices in postmenopausal women," *Osteoporosis International*, vol. 18, no. 7, pp. 963-972, 2007.

[16] C. S. Kwak, M. S. Lee, H. J. Lee, J. Y. Whang, and S. C. Park, "Dietary source of vitamin B_{12} intake and vitamin B_{12} status in female elderly Koreans aged 85 and older living in rural area," *Nutritional Research and Practice*, vol. 4, no. 3, pp. 229-234, 2010.

[17]   C. S. Kwak, M. S. Lee, S. I. Oh, and S. C. Park, "Discovery of novel sources of vitamin B_{12} in traditional Korean foods from nutritional surveys of centenarians," *Current Gerontology and Geriatrics Research*, vol. 2010, Article ID 374897, 2010.

[18]   F. Santos, A. Wegkamp, W. M. de Vos, E. J. Smid, and J. Hugenholtz, "High-level folate production in fermented foods by the B_{12} producer *Lactobacillus reuteri* JCM1112," *Applied and Environmental Microbiology*, vol. 74, no. 10, pp. 3291-3294, 2008.

[19]   J. E. Spallholz, L. M. Boylan, and J. A. Driskell, *Nutrition: Chemistry and Biology*, CRC Press, Boca Raton, FL, USA, 1999.

[20]   W. Sybesma, M. Starrenburg, M. Kleerebezem, I. Mierau, W. M. de Vos, J. Hugenholtz, "Increased production of folate by metabolic engineering of *Lactococcus lactis*," *Applied and Environmental Microbiology*, vol. 69, no. 6, pp. 3069-3076, 2003.

[21]   A. Wegkamp, M. Starrenburg, W. M. de Vos, J. Hugenholtz, and W. Sybesma, "Transformation of folate-consuming *Lactobacillus gasseri* into a folate producer," *Applied and Environmental Microbiology*, vol. 70, no. 5, pp. 3146-3148, 2004.

[22]   C. S. Yang, J. Ju, G. Lu, H. Xiao, X. Hao, S. Sang, and J. D. Lambert, "Cancer prevention by tea and tea polyphenols," *Asia Pacific Journal of Clinical Nutrition*, vol. 17, supplement 1, pp. 245-248, 2008.

[23]   T. R. Nansel, D. L. Haynie, L. M. Lipsky, L. M. B. Laffel, S. N. Mehta, "Multiple indicators of poor diet quality in children and adolescents with type 1 diabetes are associated with higher body mass index percentile but not glycemic control," *Journal of the Academy of Nutrition and Dietetics*, vol. 112, no. 11, pp. 1728-1735, 2012.

[24]   A. Drewnowski, "Concept of a nutritious food: toward a nutrient density score," *The American Journal of Clinical Nutrition*, vol. 82, no. 4, pp. 721-732, 2005.

[25]   R. Akuzawa, T. Miura, and H. Kawakami, "Bioactive components in caseins, caseinates, and cheeses," in *Bioactive Components in Milk and Dairy Products*, Y.

W. Park, Ed.,. pp. 217-233, Wiley-Blackwell, Ames, IA, USA, 2009.

[26] R. W. Hutkins, *Microbiology and Technology of Fermented Foods*, Blackwell Publishing, Ames, IA, USA, 2006.

[27] E. M. Ibeagha-Awemu, J.-R. Liu, and X. Zhao, "Bioactive components in yogurt products," in *Bioactive Components in Milk and Dairy Products*, Y. W. Park, Ed., pp. 235-250, Wiley-Blackwell, Ames, IA, USA, 2009.

[28] A. Drewnowski and N. Darmon, "The economics of obesity: dietary energy density and energy cost," *The American Journal of Clinical Nutrition*, vol. 82, no.1, pp. 265S-273S, 2005.

[29] NEPA-UNICAMP, *Tabela Brasileira de Composição de Alimentos*, NEPA-UNICAMP, Campinas, São Paulo, Brazil, 2011.

[30] N. M. O'Brien and T. P. O'Connor, "Nutritional aspects of cheese," in *Cheese; Chemistry, Physics and Microbiology. Volume 1. General Aspects*, P. F. Fox, P. L. H. McSweeney, T. M. Cogan, and T. P. Guinee, Eds., pp. 573-581, Elsevier Academic Press, London, UK, 3rd edition, 2004.

[31] R. R. Ferreira, V. A. Varisi, L. W. Meinhardt, P. J. Lea, and R. A. Azevedo, "Are high-lysine cereal crops still a challenge?," *Brazilian Journal of Medical and Biological Research*, vol. 38, no. 7, pp. 985-994, 2005.

[32] F. R. Jack and A. Paterson, "Texture of hard cheeses," *Trends in Food Science & Technology*, vol. 3, pp. 160-164, 1992.

[33] P. L. H. McSweeney, G. Ottogalli, and P. F. Fox, "Diversity of cheese varieties: an overview," in *Cheese: Chemistry, Physics and Microbiology. Volume 2. General Aspects*, P. F. Fox, P. L. H. McSweeney, T. M. Cogan, and T. P. Guinee, Eds., pp. 1-22, Elsevier Academic Press, London, UK, 3rd edition, 2004.

[34] G. Kaebnick, "Synthetic life: a new industrial revolution (2012)," in *Perspectives in Bioethics, Science, and Public Policy*, J. Beever and N. Morar, Eds., pp. 137-154, Purdue University, USA, 2013.

[35] R. T. Mahajan and G. M. Chaudhari, "Plant latex as vegetable source for milk

clotting enzymes and their use in cheese preparation," *International Journal of Advanced Research*, vol. 2, no. 5, pp. 1173-1181, 2014.

[36] *Dictionary of Food Science and Technology*, John Wiley & Sons Ltd., Chichester, West Sussex, UK, 2009.

[37] C. Spălățelu, "Biotechnological valorization of whey," *Innovative Romanian Food Biotechnology*, vol. 10, pp. 1-8, 2012.

[38] G. Barbieri, C. Barone, A. Bhagat, G. Caruso, Z. R. Conley, and S. Parisi, "The problem of aqueous absorption in processed cheeses: a simulated approach," in *The Influence of Chemistry on New Foods and Traditional Products*, pp. 1-17, Springer International Publishing, Cham, Switzerland, 2014.

[39] L. P. de S. Vandenberghe, C. T. Yamaguishi, C. Rodrigues, M. R. M. Prado, M. R. Spier, A. B. P. Medeiros, and J. C. de Carvalho, "Production of dairy products," in *Fermentation Processes Engineering in the Food Industry*, C. R. Soccol, A. Pandey, and C. Larroche, Eds., pp. 323-353, CRC Press, Boca Raton, FL, USA, 2013.

[40] J. Barłowska, M. Szwajkowska, Z. Litwińczuk, and A. Matwijczuk, "The influence of cow breed and feeding system on the dispersion state of milk fat and content of cholesterol," *Roczniki Naukowe Polskiego Towarzystwa Zootechnicznego*, vol. 7, no. 3, pp. 57-65, 2011.

[41] D. Mogul and E. Sibley, "Congenital disorders of digestion and absorption," in *Diarrhea: Diagnostic and Therapeutic Advances*, S. Guandalini and H. Vaziri, Eds., pp. 159-175, Springer Science+Business Media, LLC, New York, NY, USA, 2011.

[42] M. Temesgen and N. Ratta, "Risk of lactose intolerance and dairy food nutrition: a review," *Food Science and Quality Management*, vol. 37, pp. 19-31, 2015.

[43] R. Gandolfi, F. Gaspari, L. Franzetti, and F. Molinari, "Hydrolytic and synthetic activities of esterases and lipases of non-starter bacteria isolated from cheese surface," *Annals of Microbiology*, vol. 50, no. 2, pp. 183-189, 2000.

[44] J. Schröder, I. Maus, E. Trost, and A. Tauch, "Complete genome sequence of *Corynebacterium variabile* DSM 44702 isolated from the surface of smear-ripened cheeses and insights into cheese ripening and flavor generation," *BMC Genomics*, vol. 12, no. 545, 2011.

[45] P. Galaup, A. Gautier, Y. Piriou, A. de Villeblanche, A. Valla, and L. Dufossé, "First pigment fingerprints from the rind of French PDO red-smear ripened soft cheeses Epoisses, Mont d'Or and Maroilles," *Innovative Food Science and Emerging Technologies*, vol. 8, no. 3, pp. 373-378, 2007.

[46] G. Oboh and C. A. Elusiyan, "Changes in the nutrient and anti-nutrient content of micro-fungi fermented cassava flour produced from low- and medium-cyanide variety of cassava tubers," *African Journal of Biotechnology*, vol. 6, no. 18, pp. 2150-2157, 2007.

[47] C. E. Okpako, V. O. Ntui, A. N. Osuagwu, and F. I. Obasi, "Proximate composition and cyanide content of cassava peels fermented with *Aspergillus niger* and *Lactobacillus rhamnosus*," *Journal of Food, Agriculture and Environment*, vol. 6, no. 2, pp. 251-255, 2008.

[48] Y. Schneider, C. Kluge, U. Weiß, and H. Rohm, "Packaging materials and equipment," in *Technology of Cheesemaking*, B. A. Law and A.Y. Tamime, Eds., pp. 413-439, Wiley-Blackwell, Chichester, West Sussex, UK, 2nd edition, 2010.

[49] M. C. de Andrade Jr. and J. S. Andrade, "Physicochemical changes in cubiu fruits (*Solanum sessiliflorum* Dunal) at different ripening stages," *Food Science and Technology (Campinas)*, vol. 32, no. 2, pp. 250-254, 2012.

[50] J. D. Spence, "Intensive risk factor control in stroke prevention," *F1000Prime Reports*, vol. 5, no. 42, 2013.

[51] K. Thu, Y. Y. Mon, T. A. Khaing, and O. M. Tun, "Study on phytochemical properties, antibacterial activity and cytotoxicity of *Aloe vera* L.," *International Scholarly and Scientific Research & Innovation*, vol. 7, no. 5, pp. 114-118. 2013.

[52] N. L. Ibrahim, "Study of serum copper and iron in children with chronic liver

diseases," *Anatomy & Physiology*, vol. 4, no. 130, 2013.

[53]   A. A. Rashed, "A concise review: influence of mineral nutrients on non-haem iron uptake," *Journal of Food Technology*, vol. 10, no. 5-6, pp. 133-139, 2012.

[54]   *Oxford Dictionary of Biochemistry and Molecular Biology*, R. Cammack, Ed., Oxford University Press Inc., New York, NY, USA, 2006.

[55]   N. Scheers, L. Rossander-Hulthen, I. Torsdottir, and A.-S. Sandberg, "Increased iron bioavailability from lactic-fermented vegetables is likely an effect of promoting the formation of ferric iron (Fe^{3+})," *European Journal of Nutrition*, 2015.

[56]   H. W. Lopez, F. Leenhardt, C. Coudray, and C Remesy, "Minerals and phytic acid interactions: is it a real problem for human nutrition?," *International Journal of Food Science & Technology*, vol. 37, no. 7, pp. 727-739, 2002.

[57]   H. E. Sauberlich, *Laboratory for the Assessment of Nutritional Status*, CRC Press, Boca Raton, FL, USA, 1999.

[58]   Y. Nose, L. K. Wood, B.-E. Kim, J. R. Prohaska, R. S. Fry, J. W. Spears, and D. J. Thiele, "Ctr1 is an apical copper transporter in mammalian intestinal epithelial cells in vivo that is controlled at the level of protein stability," *The Journal of Biological Chemistry*, vol. 285, no. 42, pp. 32385-32392, 2010.

[59]   S. L. Morgan and R. L. Weinsier, *Fundamentals of Clinical Nutrition*, Mosby, St. Louis, MO, USA, 1998.

[60]   D. Bousquet-Moore, R. E. Mains, and B. A. Eipper, "PAM and copper – a gene/nutrient interaction critical to nervous system function," *Journal of Neuroscience Research*, vol. 88, 12, pp. 2535-2545, 2010.

[61]   E. T. Ojeniyi and O. O. Tewe, "Processing technology and safety of 'Kpukpuru': an indigenous weaning food in Nigeria's rural communities," *Tropical Agriculture (Trinidad and Tobago)*, vol. 75, no. 2, pp. 276-278, 1998.

[62]   M. Sotomey, E. A. D. Ategbo, E. C. Mitchikpe, M. L. Gutierrez, and M. C. Nago, *Innovations et Diffusion de Produits Alimentaires en Afrique: L'attieke au Benin*, CIRAD, Montpellier, France, 2001.

[63] A. Gegios, R. Amthor, B. Maziya-Dixon, C. Egesi, S. Mallowa, R. Nungo, S. Gichuki, A. Mbanaso, and M. J Manary, "Children consuming cassava as a staple food are at risk for inadequate zinc, iron, and vitamin A intake," *Plant Foods for Human Nutrition*, vol. 65, no. 1, pp. 64-70, 2010.

[64] S. Alam, I. Ahmad, and F. K. Bangash, "Investigation of specific elemental distribution in *C. sativas* L., *S. melongena* L. and *M. charantia* L. by atomic absorption spectrometry," *Journal of Chemical Society of Pakistan*, vol. 31, no. 1, pp. 31-38, 2009.

[65] J. M. Chiera, J. J. Finer, and E. A. Grabau, "Ectopic expression of a soybean phytase in developing seeds of *Glycine max* to improve phosphorus availability," *Plant Molecular Biology*, vol. 56, no. 6, pp. 895-904, 2004.

[66] M. Hirabayashi, T. Matsui, and H. Yano, "Fermentation of soybean meal with *Aspergillus usamii* improves zinc availability in rats," *Biological Trace Element Research*, vol. 61, no. 2, pp. 227-234, 1998.

[67] C. Hotz, "Dietary indicators for assessing the adequacy of population zinc intakes," *Food & Nutrition Bulletin*, vol. 28, supplement 3, pp. 430S-453S, 2007.

[68] Y. Motarjemi and M. J. R. Nout, "Food fermentation: a safety and nutritional assessment," *Bulletin of the World Health Organization*, vol. 74, no. 6, pp. 553-559, 1996.

[69] I. Patwary and K. Rahman, "Probiotics: should we consider it in therapeutics?," *Journal of Medicine*, vol. 13, no. 1, pp. 55-59, 2012.

[70] A. Leonardi, S. Zanoni, M. De Lucia, A. Amaretti, S. Raimondi, and M. Rossi, "Zinc uptake by lactic acid bacteria," *ISRN Biotechnology*, vol. 2013, article ID 312917, 2013.

[71] R. Jayabalan, R. V. Malbaša, E. S. Lončar, J. S. Vitas, and M. Sathishkumar, "A review on kombucha tea—microbiology, composition, fermentation, beneficial effects, toxicity, and tea fungus," *Comprehensive Reviews in Food Science and Food Safety*, vol. 13, no. 4, pp. 538-550, 2014.

[72] B. V. Mohite, G. A. Chaudhari, H. S. Ingale, and V. N. Mahajan, "Effect of fermentation and processing on in vitro mineral estimation of selected fermented foods," *International Food Research Journal*, vol. 20, no. 3, pp. 1373-1377, 2013.

[73] B. L. Kawatra and S. Kaur, "Availability of zinc from germinated, fermented and autoclaved black-gram (*Phaseolus mungo*) in rats," *Nahrung*, vol. 33, no. 4, pp. 311-314, 1989.

[74] S. T. Omaye, *Food and Nutritional Toxicology*, CRC Press, Boca Raton, FL, USA, 2004.

[75] S. P. Murphy and S. I. Barr, "Recommended Dietary Allowances should be used to set Daily Values for nutrition labeling," *The American Journal of Clinical Nutrition*, vol. 83, no. 5, pp. 1223S-1227S, 2006.

[76] P. Günther, W. Hermann, H. J. Kühn, and A. Wagner, "Wilson's disease," *Therapeutische Umschau*, vol. 64, no. 1, pp. 57-61, 2007.

[77] S. J. Moore and M. J. Warren, "The anaerobic biosynthesis of vitamin B_{12}," *Biochemical Society Transactions*, vol. 40, no. 3, pp. 581-586, 2012.

[78] M. R. D'Aimmo, P. Mattarelli, B. Biavati, N. G. Carlsson, and T. Andlid, "The potential of bifidobacteria as a source of natural folate," *Journal of Applied Microbiology*, vol. 112, pp. 975-984, 2012.

[79] F. Watanabe, Y. Yabuta, T. Bito, and F. Teng, "Vitamin B_{12}-containing plant food sources for vegetarians," *Nutrients*, vol. 6, pp. 1861-1873, 2014.

[80] S. S. Kozhakhmetov, S. S. Oralbayeva, A. R. Kushugulova, K. Kh. Almagambetov, A. B. Abzhalelov, and E. M. Ramankulov, "Creation of the probiotic consortium on the base of strains of *Bifidobacterium* spp.," *Malaysian Journal of Microbiology*, vol. 5, no. 2, pp. 67-72, 2009.

[81] K. Lagesen, D. W. Ussery, and T. M. Wassenaar, "Genome update: the 1000th genome – a cautionary tale," *Microbiology*, vol. 156, no. 3, pp. 603-608, 2010.

[82] E. E. Lioliou, A. A. Pantazaki, and D. A. Kyriakidis, "*Thermus thermophilus* genome analysis: benefits and implications," *Microbial Cell Factories*, vol. 3, no. 5.

2004.

[83] K. Liu, "Fermented soy foods: an overview," in *Handbook of Food and Beverage Fermentation Technology*, Y. H. Hui, L. Meunier-Goddik, Ö. S. Hansen, J. Josephsen, W.-K. Nip, P.S. Stanfield, and F. Toldra, Eds., Marcel Dekker, New York, NY, USA, pp. 481-495, 2004.

[84] J.-H. Martens, H. Barg, M. J. Warren, and D. Jahn, "Microbial production of vitamin B_{12}," *Applied Microbiology and Biotechnology*, vol. 58, no. 3, pp. 275-285, 2002.

[85] S. A. Survase, I. B. Bajaj, and R. S. Singhal, "Biotechnological production of vitamins," *Food Technology and Biotechnology*, vol. 44, no. 3, pp. 381-396, 2006.

[86] S. J. Moore, A. D. Lawrence, R. Biedendieck, E. Deery, S. Frank, M. J. Howard, S. E. J. Rigby, and M. J. Warren, "Elucidation of the anaerobic pathway for the corrin component of cobalamin (vitamin B_{12})," *Proceedings of the National Academy of Sciences*, vol. 110, no. 37, pp. 14906-14911, 2013.

[87] E. Hamed, M. S. Attia, and K. Bassiouny, "Synthesis, spectroscopic and thermal characterization of copper (II) and iron (III) complexes of folic acid and their absorption efficiency in the blood," *Bioinorganic Chemistry and Applications*, vol. 2009, article ID 979680, 2009.

[88] D. Oehler, A. Poehlein, A. Leimbach, N. Müller, R. Daniel, G. Gottschalk, and B. Schink, "Genome-guided analysis of physiological and morphological traits of the fermentative acetate oxidizer *Thermacetogenium phaeum*," *BMC Genomics*, vol. 13, no. 723, 2012.

[89] D. Aşlar and H. Taþtan, "Prevalence of MTHFR, MTR and MTRR gene polymorphisms in Turkish patients with nonsyndromic cleft lip and palate," *Gene Therapy and Molecular Biology*, vol. 16, pp. 115-129, 2014.

[90] L. H. Allen, "Causes of vitamin B_{12} and folate deficiency," *Food and Nutrition Bulletin*, vol. 29, no. 2, supplement, pp. S20- S34, 2008.

[91] H.-W. Chang, K.-Ho. Kim, Y.-D. Nam, S. W. Roh, M.-S. Kim, C. O. Jeon, H.-

M. Oh, and J.-W. Bae, "Analysis of yeast and archaeal population dynamics in kimchi using denaturing gradient gel electrophoresis," *International Journal of Food Microbiology*, vol. 126, no. 1-2, pp. 159-166, 2008.

[92] J. De Dea Lindner, A. L. B. Penna, I. M. Demiate, C. T. Yamaguishi, M. R. M. Prado, and J. L. Parada, "Fermented foods and human health benefits of fermented functional foods," in *Fermentation Processes Engineering in the Food Industry*, C. R. Soccol, A. Pandey, and C. Larroche, Eds., CRC Press, Boca Raton, FL, USA, 2013.

[93] S. C. Park, "Comprehensive approach for studying longevity in Korean centenarians," *Asian Journal of Gerontology & Geriatrics*, vol. 7, no. 1, pp. 33-38, 2012.

[94] K.-G. Lee, "Analysis and risk assessment of ethyl carbamate in various fermented foods," *European Food Research and Technology*, vol. 236, no. 5, pp. 891-898, 2013.

[95] M. Egger, S. Mutschlechner, N. Wopfner, G. Gadermaier, P. Briza, and F. Ferreira, "Pollen-food syndromes associated with weed pollinosis: an update from the molecular point of view," *Allergy*, vol. 61, no. 4, pp. 461-476, 2006.

[96] H.-B. Jang, Y.-H. Han, C. J. Piyathilake, H. Kim, and T. Hyun, "Intake and blood concentrations of folate and their association with health-related behaviors in Korean college students," *Nutrition Research and Practice*, vol. 7, no. 3, pp. 216-223, 2013.

[97] R. B. Stewart and A. Getaciiew, "Investigations of the nature of Injera," *Economic Botany*, vol. 16, no. 2, pp. 127-130, 1962.

[98] A.-S. Hager, F. Lauck, E. Zannini, and E. K. Arendt, "Development of gluten-free fresh egg pasta based on oat and teff flour," *European Food Research and Technology*, vol. 235, no. 5, pp. 861-871, 2012.

[99] S. Khokhar, O. J. Oyelade, L. Marletta, D. Shahar, J. Ireland, and S. de Henauw, "Vitamin composition of ethnic foods commonly consumed in Europe," *Food & Nutrition Research*, vol. 56, no. 5639, 2012.

[100] E. Gujska and K. Majewska, "Effect of baking process on added folic acid and endogenous folates stability in wheat and rye breads," *Plant Foods for Human Nutrition*, vol. 60, no. 2, pp 37-42, 2005.

[101] R. López-Nicolás, C. Frontela-Saseta, R. González-Abellán, A. Barado-Piqueras, D. Perez-Conesa, and G. Ros-Berruezo, "Folate fortification of white and whole-grain bread by adding Swiss chard and spinach. Acceptability by consumers," *LWT – Food Science and Technology*, vol. 59, no. 1, pp. 263-269, 2014.

[102] T. B. Fitzpatrick, G. J. C. Basset, P. Borel, F. Carrari, D. DellaPenna, P. D. Fraser, H. Hellmann, S. Osorio, C. Rothan, V. Valpuesta, C. Caris-Veyrat, and A. R. Fernie, "Vitamin deficiencies in humans: can plant science help?," *The Plant Cell*, vol. 24, no. 2, pp. 395-414, 2012.

[103] S. Tibbott, "Tempeh: the ''other'' white beancake," in *Handbook of Food and Beverage Fermentation Technology*, Y. H. Hui, L. Meunier-Goddik, Ö. S. Hansen, J. Josephsen, W.-K. Nip, P. S. Stanfield, and F. Toldra, Eds., pp. 583-594, Marcel Dekker, New York, NY, USA, 2004.

[104] G. Lippi and M. Franchini, "Vitamin K in neonates: facts and myths," *Blood Transfusion*, vol. 9, no. 1, pp. 4-9, 2011.

[105] J. T. Truong and S. L. Booth, "Emerging issues in vitamin K research," *Journal of Evidence-Based Complementary & Alternative Medicine*, vol. 16, no. 1, pp. 73-79, 2011.

[106] H. Rasekhi, M. Karandish, M. T. Jalali, M. Mohammadshahi, M. Zarei, A. Saki, H. Shahbazian, "Phylloquinone supplementation improves glycemic status independent of the effects of adiponectin levels in premonopause women with prediabetes: a double-blind randomized controlled clinical trial," *Journal of Diabetes & Metabolic Disorders*, vol. 14, no. 1, 2015.

[107] S. L. Booth and J. W. Suttie, "Dietary intake and adequacy of vitamin K," The *Journal of Nutrition*, vol. 128, no. 5, pp.785-788, 1998.

[108] S. Truswell, "Vitamins D and K," in *Essentials of Human Nutrition*, J. Mann

and A. S. Truswell, Eds., pp. 254-258, Oxford University Press, New York, NY, USA, 2002.

[109] C. Vermeer, "Vitamin K: the effect on health beyond coagulation an overview," *Food & Nutrition Research*, vol. 56, no. 5329, 2012.

[110] S. M. Cutting, "*Bacillus* probiotics," *Food Microbiology*, vol. 28, no. 2, pp. 214-220, 2011.

[111] A. Shimakage, M. Shinbo, and S. Yamada, "ACE inhibitory substances derived from soy foods," *Journal of Biological Macromolecules*, vol. 12, no. 3, pp. 72-80, 2012.

[112] M. Muguruma, A. M. Ahhmed, S. Kawahara, K. Kusumegi, T. Hishinuma, K. Ohya, and T. Nakamura, "A combination of soybean and skimmed milk reduces osteoporosis in rats," *Journal of Functional Foods*, vol. 4, no. 4, pp. 810-818, 2012.

[113] S. L. Booth, "Vitamin K: food composition and dietary intakes," *Food & Nutrition Research*, vol. 56, no. 5505, 2012.

[114] N. Gad and M. R. Abd El-Moez, "Broccoli growth, yield quantity and quality as affected by cobalt nutrition," *Agriculture and Biology Journal of North America*, vol. 2, no. 2, pp. 226-231, 2011.

[115] J. Conly, J. Suttie, E. Reid, J. Loftson, K. Ramotar, and T. Louie, "Dietary deficiency of phylloquinone and reduced serum levels in febrile neutropenic cancer patients," *The American Journal of Clinical Nutrition*, vol. 50, no. 1, pp. 109-113, 1989.

[116] H. M. Macdonald, F. E. McGuigan, S. A. Lanham-New, W. D. Fraser, S. H. Ralston, and D. M. Reid, "Vitamin K_1 intake is associated with higher bone mineral density and reduced bone resorption in early postmenopausal Scottish women: no evidence of gene-nutrient interaction with apolipoprotein E polymorphisms," *The American Journal of Clinical Nutrition*, vol. 87, no. 5, pp. 1513-1520, 2008.

[117] K. Nimptsch, S. Rohrmann, and J. Linseisen, "Dietary intake of vitamin K and risk of prostate cancer in the Heidelberg cohort of the European Prospective

Investigation into Cancer and Nutrition (EPIC-Heidelberg)," *The American Journal of Clinical Nutrition*, vol. 87, no. 4, pp. 985-992, 2008.

[118] G. J. Tortora, B. R. Funke, and C. L. Case, *Microbiology: an Introduction*, Benjamin Cummings, San Francisco, CA, USA, 2010.

[119] Y. Funakoshi, H. Omori, and T. Katoh, "Association between bone mineral density and lifestyle factors or vitamin D receptor gene polymorphism in adult male workers: a cross-sectional study," *Environmental Health and Preventive Medicine*, vol. 14, no. 6, pp. 328-335, 2009.

[120] A. Miloš, A. Selmanović, L. Smajlović, R. L. M. Huel, C. Katzmarzyk, A. Rizvić, and T. J. Parsons, "Success rates of nuclear short tandem repeat typing from different skeletal elements," *Croatian Medical Journal*, vol. 48, pp. 486-493, 2007.

[121] P.-S. Juo, *Concise Dictionary of Biomedicine and Molecular Biology*, CRC Press, Boca Raton, FL, USA, 2002.

[122] M. J. Seibel and C. Meier, "Biochemical markers of bone turnover – basic biochemistry and variability," in *Osteoporosis: Pathophysiology and Clinical Management*, R. A. Adler, Ed., pp. 97-130, Humana Press, New York, NY, USA, 2010.

[123] K. D. Cashman, "Diet, nutrition, and bone health," *The Journal of Nutrition*, vol. 137, no. 11, pp. 2507S-2512S, 2007.

[124] V. Schwetz, T. Pieber, and B. Obermayer-Pietsch, "The endocrine role of the skeleton: background and clinical evidence," *European Journal of Endocrinology*, vol. 166, no. 6, pp. 959-967, 2012.

[125] A. Patti, L. Gennari, D. Merlotti, F. Dotta, and R. Nuti, "Endocrine actions of osteocalcin," *International Journal of Endocrinology*, vol. 2013, article ID 846480, 2013.

[126] C. B. Confavreux, "Bone: from a reservoir of minerals to a regulator of energy metabolism," *Kidney International*, vol. 79, supplement 121, S14-S19, 2011.

[127] K. S. Gravenstein, J. K. Napora, R. G. Short, R. Ramachandran, O. D. Carlson,

E. J. Metter, L. Ferrucci, J. M. Egan, and C. W. Chia, "Cross-sectional evidence of a signaling pathway from bone homeostasis to glucose metabolism," *The Journal of Clinical Endocrinology & Metabolism*, vol. 96, no. 6, pp. E884-E890, 2011.

[128] M. Yamaguchi, "New development in osteoporosis treatment: the synergistical osteogenic effects with vitamin D_3, menaquinone-7, genistein and zinc," *Vitamins and Minerals*, vol. S6: e001, 2013.

[129] I. Al-Saleha, N. Shinwaria, A. Mashhour, G. E.-D. Mohamed, M. A. Ghosh, Z. Shammasi, and A. Al-Nasser, "Is lead considered as a risk factor for high blood pressure during menopause period among Saudi women?," *International Journal of Hygiene and Environmental Health*, vol. 208, no. 5, pp. 341-356, 2005.

[130] Y. Fang, J. B. J. van Meurs, P. Arp, J. P. T. van Leeuwen, A. Hofman, H. A. P. Pols, and A. G. Uitterlinden, "Vitamin D binding protein genotype and osteoporosis," *Calcified Tissue International*, vol. 85, no. 2, pp. 85-93, 2009.

[131] A. S. Issever, T. M. Link, M. Kentenich, P. Rogalla, A. J. Burghardt, G. J. Kazakia, S. Majumdar, and G. Diederichs, "Assessment of trabecular bone structure using MDCT: comparison of 64- and 320-slice CT using HR-pQCT as the reference standard," *European Radiology*, vol. 20, no. 2, pp. 458-468, 2010.

[132] J. A. Kanis, N. Burlet, C. Cooper, P. D. Delmas, J.-Y. Reginster, F. Borgstrom, and R. Rizzoli, "European guidance for the diagnosis and management of osteoporosis in postmenopausal women," *Osteoporosis International*, vol. 19, no. 4, pp. 399-428, 2008.

[133] Meeta, C. V. Harinarayan, R. Marwah, R. Sahay, S. Kalra, and S. Babhulkar "Clinical practice guidelines on postmenopausal osteoporosis: an executive summary and recommendations," *Journal of Mid-Life Health*, vol. 4, 2, pp.107-126, 2013.

[134] S. Katz, "Prevention, detection, and treatment of osteopenia and osteoporosis," *Gastroenterology & Hepatology*, vol. 9, no. 3, pp. 176-178, 2013.

[135] V. M. Brandenburg, L. J. Schurgers, N. Kaesler, K. Püsche, R. H. van Gorp, G. Leftheriotis, S. Reinartz, R. Koos, and T. Krüger, "Prevention of vasculopathy by

vitamin K supplementation: can we turn fiction into fact?," *Atherosclerosis*, vol. 240, no. 1, pp. 10-16, 2015.

[136] A.-M. Boudet, "Evolution and current status of research in phenolic compounds," *Phytochemistry*, vol. 68, no. 22-24, pp. 2722-2735, 2007.

[137] A. Farah and C.M. Donangelo, "Phenolic compounds in coffee," *Brazilian Journal of Plant Physiology*, vol. 18, no. 1, pp. 23-36, 2006.

[138] T. Shoji, Y. Akazome, T. Kanda, and M. Ikeda, "The toxicology and safety of apple polyphenol extract," *Food and Chemical Toxicology*, vol. 42, no. 6, pp. 959-967, 2004.

[139] P. Delaveau, "Polyphénols et tanins dans l'alimentation," *Cahiers de Nutrition et de Diététique*, vol. 23, no. 2, pp. 137-139, 1988.

[140] J. H. M. Henderson, "Fifty years as a plant physiologist," *Annual Review of Plant Physiology and Plant Molecular Biology*, vol. 52, pp. 1-28, 2001.

[141] E. Ortega, M. C. Sadaba, A. I. Ortiz, C. Cespon, A. Rocamora, J. M. Escolano, G. Roy, L. M. Villar, and P. Gonzalez-Porque, "Tumoricidal activity of lauryl gallate towards chemically induced skin tumours in mice," *British Journal of Cancer*, vol. 88, pp. 940-943, 2003.

[142] K. B. Pandey and S. I. Rizvi, "Plant polyphenols as dietary antioxidants in human health and disease," *Oxidative Medicine and Cellular Longevity*, vol. 2, no. 5, pp. 270-278, 2009.

[143] S. Cosmulescu, I. Trandafir, G. Achim, M. Botu, A. Baciu, and M. Gruia, "Phenolics of green husk in mature walnut fruits," *Notulae Botanicae Horti Agrobotanici Cluj-Napoca*, vol. 38, no. 1, pp. 53-56, 2010.

[144] J. Dai and R. J. Mumper, "Plant phenolics: extraction, analysis and their antioxidant and anticancer properties," *Molecules*, vol. 15, no. 10, pp. 7313-7352, 2010.

[145] A. García-Lafuente, E. Guillamón, A. Villares, M. A. Rostagno, and J. A. Martínez, "Flavonoids as anti-inflammatory agents: implications in cancer and

cardiovascular disease," *Inflammation Research*, vol. 58, no. 9, pp. 537-552, 2009.

[146] K. B. Pandey and S. I. Rizvi, "Current understanding of dietary polyphenols and their role in health and disease," *Current Nutrition & Food Science*, vol. 5, no. 4, pp. 249-263, 2009.

[147] M. Rossi and A. Amaretti, "Probiotic properties of bifidobacteria," in *Bifidobacteria: Genomics and Molecular Aspects*, B. Mayo and D. van Sinderen, Eds., pp. 97-124, Caister Academic Press, Norfolk, UK,. 2010.

[148] M. Rusconi and A. Conti, "*Theobroma cacao* L., the food of the gods: a scientific approach beyond myths and claims," *Pharmacological Research*, vol. 61, no. 1, pp. 5-13, 2010.

[149] D. Vauzour, K. Vafeiadou, A. Rodriguez-Mateos, C. Rendeiro, and J. P. E. Spencer, "The neuroprotective potential of flavonoids: a multiplicity of effects," *Genes & Nutrition*, vol. 3, no. 3-4, pp. 115-126, 2008.

[150] C. Corredor, T. Teslova, M. V. Canãmares, C. Zhanguo, J. Zhang, and J. R. Lombardi, "Raman and surface-enhanced Raman spectra of chrysin, apigenin and luteolin," *Vibrational Spectroscopy*, vol. 49, no. 2, pp. 190-195, 2009.

[151] P. Brasnyo, G. A. Molnár, M. Mohás, L. Markó, B. Laczy, J. Cseh, E. Mikolás, I. A. Szijártó, A. Mérei, R. Halmai, L. G. Mészáros, B. Sümegi, and I. Wittmann, "Resveratrol improves insulin sensitivity, reduces oxidative stress and activates the Akt pathway in type 2 diabetic patients," *British Journal of Nutrition*, vol. 106, no. 3, pp 383-389, 2011.

[152] N. J. Kang, K. W. Lee, J. Y. Kwon, M. K. Hwang, E. A. Rogozin, Y.-S. Heo, A. M. Bode, H. J. Lee, and Z. Dong, "Delphinidin attenuates neoplastic transformation in JB6 CL41 mouse epidermal cells by blocking Raf/mitogen-activated protein kinase kinase/extracellular signal-regulated kinase signaling," *Cancer Prevention Research*, vol. 1, no. 7, pp. 522-531, 2008.

[153] C. G. Fraga, M. C. Litterio, P. D. Prince, V. Calabró, B. Piotrkowski, and M. Galleano, "Cocoa flavanols: effects on vascular nitric oxide and blood pressure,"

Journal of Clinical Biochemistry and Nutrition, vol. 48, no. 1, pp. 63-67, 2011.

[154] J. E. Lozano, *Fruit Manufacturing: Scientific Basis, Engineering Properties, and Deteriorative Reactions of Technological Importance*, Springer Science, New York, NY, USA, 2006.

[155] J. Shi, S. J. Xue, and Y. Kakuda, "Green tea-induced thermogenesis controlling body weight," in *Tea and Tea Products: Chemistry and Health-Promoting Properties*, C.-T. Ho, J.-K. Lin, and F. Shahidi, Eds., pp. 221-232, CRC Press, Boca Raton, FL, USA, 2009.

[156] T. Tanaka, K. Inoue, Y. Betsumiya, C. Mine, and I. Kouno, "Two types of oxidative dimerization of the black tea polyphenol theaflavin," Journal of Agricultural and Food Chemistry, vol. 49, no. 12, pp. 5785-5789, 2001.

[157] I. M. Tabu, V. M. Kekana, and D. M. Kamau, "Effect of varying ratios and rates of enriched cattle manure on leaf nitrogen content, yield and quality of tea *(Camellia sinensis)*," *Journal of Agricultural Science*; vol. 7, no. 5, pp. 175-181 2015.

[158] N. N. Tram, P. P. Hien, and H. N. Oanh, "Change of polyphenol oxidase activity during oolong tea process," *Journal of Food and Nutrition Sciences*, vol. 3, no. 1-2, pp. 88-93, 2015.

[159] A. Gomes, P. Datta, A. Sarkar, S. C. Dasgupta, and A. Gomes, "Black tea *(Camellia sinensis)* extract as an immunomodulator against immunocompetent and immunodeficient experimental rodents," *Oriental Pharmacy and Experimental Medicine*, vol. 14, no. 1, pp 37-45, 2014.

[160] E. Haasbach, C. Hartmayer, A. Hettler, A. Sarnecka, U. Wulle, C. Ehrhardt, S. Ludwig, and O. Planz, "Antiviral activity of Ladania067, an extract from wild black currant leaves against influenza A virus *in vitro* and *in vivo*," *Frontiers in Microbiology*, vol. 5, no. 71, pp. 1-11, 2014.

[161] J. A. M. Kyle and G. G. Duthie, "Flavonoids in foods," in *Flavonoids: Chemistry, Biochemistry, and Applications*, Ø. M. Andersen and K. R. Markham,

Eds., pp. 219-262, CRC Press, Boca Raton, FL, USA, 2006.

[162] E. Theodoratou, J. Kyle, R. Cetnarskyj, S. M. Farrington, A. Tenesa, R. Barnetson, M. Porteous, M. Dunlop, and H. Campbell, "Dietary flavonoids and the risk of colorectal cancer," *Cancer Epidemiology, Biomarkers & Prevention*, vol. 16, no. 4, pp. 684-693, 2007.

[163] T. Baysal and A. Demirdöven, "Lipoxygenase in fruits and vegetables: a review," *Enzyme and Microbial Technology*, vol. 40, no. 4, pp. 491-496, 2007.

[164] N. Garti, G. Agmon, and E. Pintus, Method for selectively obtaining antioxidant rich extracts from citrus fruits. Patent US Pat. 6, 528, 099, 2003.

[165] G. Mishra, S. Srivastava, and B. P. Nagori, "Pharmacological and therapeutic activity of Cissus quadrangularis: an overview," *International Journal of PharmTech Research*, vol. 2, no. 2, 1298-1310, 2010.

[166] S. A. Aherne and N. M. O'Brien, "Dietary flavonols: chemistry, food content, and metabolism," *Nutrition*, vol. 18, no. 1, pp. 75-81, 2002.

[167] N. Andarwulan, R. Batari, D. A. Sandrasari, B. Bolling, and H. Wijaya, "Flavonoid content and antioxidant activity of vegetables from Indonesia," *Food Chemistry*, vol. 121, no. 4, pp. 1231-1235, 2010.

[168] A. G. Cunha, *Dicionário Etimológico Nova Fronteira da Língua Portuguesa*, Nova Fronteira, Rio de Janeiro, RJ, Brazil, 1991.

[169] I. Kubo, N. Masuokaa, K. Niheia, and B. Burgheima, "Maniçoba, a quercetin-rich Amazonian dish," *Journal of Food Composition and Analysis*, vol. 19, no. 6-7, pp. 579-588, 2006.

[170] J. A. M. Kyle, L. Sharp, J. Little, G. G. Duthie, and G. McNeill, "Dietary flavonoid intake and colorectal cancer: a case – control study," *British Journal of Nutrition*, vol. 103, no. 3, pp. 429-436, 2010.

[171] D. C. Nieman and N. C. Bishop, "Nutritional strategies to counter stress to the immune system in athletes, with special reference to football," *Journal of Sports Sciences*, vol. 24, no. 7, pp. 763-772, 2006.

[172] M. Schuier, H. Sies, B. Illek, and H. Fischer, "Cocoa-related flavonoids inhibit CFTR-mediated chloride transport across T84 human colon epithelia," *The Journal of Nutrition*, vol. 135, no. 10, pp. 2320-2325, 2005.

[173] R. H. Gokani, M. A. Rachchh, T. P. Patel, S. K. Lahiri, D. D. Santani, and M. B. Shah, "Evaluation of anti-oxidant activity (in vitro) of *Clerodendrum phlomidis*, Linn. F. suppl. Root," *Journal of Herbal Medicine and Toxicology*, vol. 5, no. 1, pp. 47-53, 2011.

[174] R. Pourabbasa, A. Delazarb, and M. T. Chitsaza, "The effect of German chamomile mouthwash on dental plaque and gingival inflammation," *Iranian Journal of Pharmaceutical Research*, vol. 4, no. 2, pp. 105-109, 2005.

[175] J. M. Hamilton-Reeves, G. Vazquez, S. J. Duval, W. R. Phipps, M. S. Kurzer, and M. J. Messina, "Clinical studies show no effects of soy protein or isoflavones on reproductive hormones in men: results of a meta-analysis," *Fertility and Sterility*, vol. 94, no. 3, pp. 997-1007, 2010.

[176] J. M. Pavese, R. L. Farmer, and R. C. Bergan, "Inhibition of cancer cell invasion and metastasis by genistein," *Cancer and Metastasis Reviews*, vol. 29, no. 3, pp. 465-482, 2010.

[177] S. Agrawal, Modulation of oligonucleotide CpG-mediated immune stimulation by positional modification of nucleosides. Patent US 0059067 A1, 2011.

[178] G. Nardone, D. Compare, P. de Colibus, G. de Nucci, and A. Rocco, "*Helicobacter pylori* and epigenetic mechanisms underlying gastric carcinogenesis," *Digestive Diseases* (Basel, Switzerland), vol. 25, no. 3, pp. 225-229, 2007.

[179] S.-W. Choi and S. Friso, "Epigenetics: a new bridge between nutrition and health," *Advances in Nutrition*, vol. 1, pp. 8-16, 2010.

[180] M. Narajczyk, M. Moskot, and A. Konieczna, "Quantitative estimation of lysosomal storage in mucopolysaccharidoses by electron microscopy analysis," *Acta Biochimica Polonica*, vol. 59, no. 4, pp. 693-696, 2012.

[181] S. Banerjee and P. Rajamani, "Cellular, molecular, and biological perspective

of polyphenols in chemoprevention and therapeutic adjunct in cancer," in *Natural Products*, K. G. Ramawat and J.-M. Mérillon, Eds., pp 2175-2254, Springer, Berlin, Germany, 2013.

[182] S.-E. Jang, K.-A. Kim, M. J. Han, and D.-H. Kim, "*Doenjang*, a fermented Korean soybean paste, inhibits lipopolysaccharide production of gut microbiota in mice," *Journal of Medicinal Food*, vol. 17, no. 1, pp. 67-75, 2014.

[183] J. Zupancic, "Probiotic use in neonates," *Nursing for Women's Health*, vol. 13, no. 1), pp. 59-64, 2009.

[184] R. Bibiloni, C. Lay, and G. W. Tannock, "Probiotics: lessons learned from nucleic acid-based analysis of bowel communities," in *Molecular Techniques in the Microbial Ecology of Fermented Foods*, L. Cocolin and D. Ercolini, Eds., pp. 225-244, Springer, New York, NY, USA, 2008.

[185] E. R. "Farnworth, Kefir – a complex probiotic," *Food Science and Technology Bulletin*, vol. 2, no. 1, pp. 1-17, 2005.

[186] P. Gionchetti, F. Rizzello, U. Helwig, A. Venturi, K. M. Lammers, P. Brigidi, B. Vitali, G. Poggioli, M. Miglioli, and M. Campieri, "Prophylaxis of pouchitis onset with probiotic therapy: a double-blind, placebo-controlled trial," *Gastroenterology*, vol. 124, pp. 1202-1209, 2003.

[187] S. Hogg, *Essential Microbiology*, John Wiley & Sons, Chichester, West Sussex, UK, 2005.

[188] E. M. Ibeagha-Awemu, J.-R. Liu, and X. Zhao, "Bioactive components in yogurt products," in *Bioactive Components in Milk and Dairy Products*, Y. W. Park, Ed., pp. 235-250, Wiley-Blackwell, Ames, Iowa, USA, 2009.

[189] J. M. Jay, M. J. Loessner, and D. A. Golden, *Modern Food Microbiology*, Springer, New York, NY, USA, 2005.

[190] S. Michail, "The role of probiotics in allergic diseases," *Allergy, Asthma & Clinical Immunology*, vol. 5, no. 5, 2009.

[191] M. T. Ukeyima, V. N. Enujiugha, and T. A. Sanni, "Current applications of

probiotic foods in Africa," *African Journal of Biotechnology*, vol. 9, no. 4, pp. 394-401, 2010.

192] M. Hayta and B. Polat, "Incorporation of nutraceutical ingredients in baked goods," in *Nutraceutical and Functional Food Processing technology*," J. E. Boye, Ed., pp. 211-234 John Wiley & Sons, Chichester, West Sussex, UK, 2015.

[193] A. M. Mahasneh and M. M. Abbas, "Probiotics and traditional fermented foods: the eternal connection (mini-review)," *Jordan Journal of Biological Sciences*, vol. 3, no. 4, pp. pp. 133-140, 2010.

[194] T. Beresford and A. Williams, "The microbiology of cheese ripening," in *Cheese Chemistry, Physics and Microbiology. Volume 1 General Aspects*, P. F. Fox, P. L.H. McSweeney, T. M. Coganpp, and T. P. Guinee, Eds., pp. 287-317, Elsevier Academic Press, San Diego, CA, USA, 2004.

[195] O. Yerlikaya and E. Ozer, "Production of probiotic fresh white cheese using co-culture with *Streptococcus thermophilus*," *Food Science and Technology* (Campinas), vol. 34, no. 3, pp. 471-477, 2014.

[196] J. C. R. Nogueira and M. D. C. R. Gonçalves, "Probiotics in allergic rhinitis," *Brazilian Journal of Otorhinolaryngology*, vol. 77, no. 1, pp. 129-134, 2011.

[197] L. Ruiz, P. Ruas-Madiedo, M. Gueimonde, C. G. de los Reyes-Gavilán, A. Margolles, and B. Sánchez, "How do bifidobacteria counteract environmental challenges? Mechanisms involved and physiological consequences," *Genes & Nutrition*, vol. 6, no. 3, pp. 307-318, 2011.

[198] L. Guerin-Deremaux, F. Ringard, F. Desailly, and D. Wils, "Effects of a soluble dietary fibre NUTRIOSE® on colonic fermentation and excretion rates in rats," *Nutrition Research and Practice*, vol. 4, no. 6, pp. 470-476, 2010.

[199] E. G. Kiarie, B. A. Slominski, D. O. Krause, and C. M. Nyachoti, "Nonstarch polysaccharide hydrolysis products of soybean and canola meal protect against enterotoxigenic *Escherichia coli* in piglets," *The Journal of Nutrition*, vol. 138, no. 3, pp. 502-508, 2008.

[200] L.-G. Ooi and M.-T. Liong, "Cholesterol-lowering effects of probiotics and prebiotics: a review of in vivo and in vitro findings," *International Journal of Molecular Sciences*, vol. 11, no. 6, pp. 2499-2522, 2010.

[201] Ś. Dominika, N. Arjan, R. P. Karyn, and K. Henryk, "The study on the impact of glycated pea proteins on human intestinal bacteria," *International Journal of Food Microbiology*, vol. 145, no. 1, pp. 267-272, 2011.

[202] S.C. Choi, B. J. Kim, P.-L. Rhee, D. K. Chang, H. J. Son, J. J. Kim, J. C. Rhee, S. I. Kim, Y. S. Han, K. H. Sim, and S. N. Park, "Probiotic fermented milk containing dietary fiber has additive effects in IBS with constipation compared to plain probiotic fermented milk," *Gut Liver*, vol. 5, no. 1, pp. 22-28, 2011.

[203] P. Guilloteau, L. Martin, V. Eeckhaut, R. Ducatelle, R. Zabielski, and F. Van Immerseel, "From the gut to the peripheral tissues: the multiple effects of butyrate," *Nutrition Research Reviews*, vol. 23, no. 2, pp. 366-384, 2010.

[204] R. B. Canani, M. Di Costanzo, L. Leone, M. Pedata, R. Meli, and A. Calignano, "Potential beneficial effects of butyrate in intestinal and extraintestinal diseases," *World Journal of Gastroenterology*, vol. 17, no. 12, pp. 1519-1528, 2011.

[205] S. A. Ross and J.-M Ekoé, "Incretin agents in type 2 diabetes," *Canadian Family Physician*, vol. 56, no. 7, pp. 639-648, 2010.

[206] J. M. Ellis, D. S. Paul, M. A. DePetrillo, B. P. Singh, D. E. Malarkey, and R. A. Coleman, "Mice deficient in glycerol-3-phosphate acyltransferase-1 have a reduced susceptibility to liver cancer," *Toxicologic Pathology*, vol. 40, pp. 513-521, 2012.

[207] L. Zhu, S. S. Baker, C. Gill, W. Liu, R. Alkhouri, R. D. Baker, and S. R. Gill, "Characterization of gut microbiomes in nonalcoholic steatohepatitis (NASH) patients: a connection between endogenous alcohol and NASH," *Hepatology*, vol. 57, no. 2, pp. 601-609, 2013.

[208] G. Sagratini, M. Fernández-Franzón, F. De Berardinis, G. Font, S. Vittori, and J. Mañes, "Simultaneous determination of eight underivatised biogenic amines in fish

by solid phase extraction and liquid chromatography–tandem mass spectrometry," *Food Chemistry*, vol. 132, no. 1, pp. 537-543, 2012.

[209] E. De Mey, G. Drabik-Markiewicz, H. De Maere, M.-C. Peeters, G. Derdelinckx, H. Paelinck, and T. Kowalska, "Dabsyl derivatisation as an alternative for dansylation in the detection of biogenic amines in fermented meat products by reversed phase high performance liquid chromatography," *Food Chemistry*, vol. 130, no. 4, pp. 1017-1023, 2012.

[210] M. A. M. AL-Ashmawy, M. M. Gwida, and K. H. Abdelgalil, "Prevalence, detection methods and antimicrobial susceptibility of *Listeria monocytogens* isolated from milk and soft cheeses and its zoonotic importance," *World Applied Sciences Journal*, vol. 29, no. 7, pp. 869-878, 2014.

[211] B. J. Silk, K. A. Date, K. A. Jackson, R. Pouillot, K. G. Holt, L. M. Graves, K. L. Ong, S. Hurd, R. Meyer, R. Marcus, B. Shiferaw, D. M. Norton, C. Medus, S. M. Zansky, A. B. Cronquist, O. L. Henao, T. F. Jones, D. J. Vugia, M. M. Farley, and B. E. Mahon, "Invasive listeriosis in the foodborne diseases active surveillance network (foodnet), 2004–2009: further targeted prevention needed for higher-risk groups," *Clinical Infectious Diseases*, vol. 54, no. supplement 5, pp. S396-S404, 2012.

[212] L. A. Chiou, T. W. Hennessy, A. Horn, G. Carter, and J. C. Butler, "Botulism among Alaska natives in the Bristol Bay area of southwest Alaska: a survey of knowledge, attitudes, and practices related to fermented foods known to cause botulism," *International Journal of Circumpolar Health*, vol. 61, no. 1, pp. 50-60, 2002.

[213] R. Chaudhry, "Botulism: a diagnostic challenge," *Indian Journal of Medical Research*, vol.134, no. 1, pp. 10-12, 2011.

[214] N. N. Potter and J. H. Hotchkiss, *Food Science*, An Aspen Publication, Gaithersburg, MD, USA, 1998.

[215] R. Rolle and M. Satin, "Basic requirements for the transfer of fermentation technologies to developing countries," *International Journal of Food Microbiology*,

vol. 75, no. 3, pp. 181-187, 2002.

CHAPTER 5

THE SOCIOECONOMIC RELEVANCE OF FERMENTED FOODS IN THE PRESENT HUMAN DIETS: THE EMPHASIS ON THE ETHNIC FERMENTED FOODS OF THE BRAZILIAN AMAZONIA

As previously emphasized, fermented foods are at the crossroads of ethnicity, culture and nutrition. In other words, national (or racial groups) and their characteristic habits, traditions and beliefs can significantly modify food intake and more specifically food choice and have strong historical antecedents, rooted in unique combinations of environment (geography, climate, plant and animal species), ritualistic systems, community and family structures, human endeavor, mobility (migrations, exploration) and economic and political systems [1-3]. These multiple factors are naturally integrated into traditional rules of cuisine and appropriateness and are then embodied in the different ethnic foods themselves [2, 4].

Ethnic foods, seldom referred to as exotic foods, have arisen much interest during the last few years, but they lack of a satisfactory definition [5, 6]. The authors of this book propose the following generic definition of ethnic foods: typical foods reminiscent of a people (or a culturally distinct group of people), usually prepared with the natural resources of this people's habitat by a simple but well-established food technology such as cooking and fermentation. Nevertheless, cultural aspects closely approach the meanings of ethnic foods and traditional foods to the extent that these expressions are used interchangeably in the literature [7-9].

In developing countries like Nigeria, fermentation techniques are passed on as trade secrets in the families of certain communities (a practice protected by tradition) and fermented foods constitute a significant component of the diets, mostly in the rural areas [10]. This African context of small-scale production of fermented foods is applicable to the vast territory of the Brazilian Amazonia. Nevertheless, diverse fermented dairy products are consumed in the urban conglomerates of this region.

However, the Indians remain active throughout the Amazonian region, where they have not been eliminated, and the less urbanized is the region; the greater is their cultural influence on other natives [11].

In the Brazilian Amazonia, fermentation is used as the main process in the production of some ethnic foods, or as a supplementary process in the production of other foods not classified as typical fermented foods (in this case, fermented products would serve as flavoring fermented ingredients in the elaboration of other foodstuffs).

As mentioned in the introduction of this book, many Amazonian fruits serve as substrates for alcoholic fermentation and alcoholic fermented beverages such as *caiçuma*, *tiquira* and *tarubá* (words of Tupi origin) [12]. Fermentation may develop sensorial attributes in *aluá* (Arabic origin), introduced in Brazil by African slaves in the sixteenth century [12]. The sensory properties of camu-camu juice may be enriched by fermentation too. The structure of andiroba seeds (*Carapa guianensis* Aubl., Meliaceae) may be altered so that its perfumed oil may be extracted. Fermentation may also be used to promote hydrolysis of pectin in order to simplify osmotic dehydration of various fruits as well as of cassava flour for sensorial enrichment.

Tucupi is in fact a liquid extracted from bitter cassava and used as a sauce in the regional cuisine of the north of Brazil (Amazonas and Pará) [13]. However, fermentation improves *tucupi* sensory characteristics, explaining its use as a delicious condiment of many recipes.

Finally, fermentation develops sensory properties in cassava gum and is used in an appreciated culinary preparation called *tacacá*. Fermentation forms flavor precursors of chocolate made of the seeds of cupuaçu. Fermentation softens buriti (*Mauritia flexuosa* L. f., Arecaceae) for its pulping.

REFERENCES

[1]   D. J. Hoffman and A. L. Sawaya, "Energy balance," in *Encyclopedia of Human Nutrition*, M. J. Sadler, J. J. Strain, and B. Caballero, Eds., pp. 650-658, Academic Press, San Diego, USA, 1999.

[2]   R. Shepherd and D. Mela, "Factors influencing food choice," in *Encyclopedia of Human Nutrition*, M. J. Sadler, J. J. Strain, and B. Caballero, Eds., pp. 843-850, Academic Press, San Diego, USA, 1999.

[3]   P. J. Brown, "Cultural perspectives on the etiology and treatment of obesity," in *Obesity: Theory and Therapy*, A. J. Stunkard and T. A. Wadden, Eds., pp. 179-193, Lippincott-Raven, Philadelphia, USA, 2nd edition, 1993.

[4]   D. J. Mela, "Food choice and intake: the human factor," *Proceedings of the Nutrition Society*, vol. 58, 513-521, 1999.

[5]   W. Verbeke and G. P. López, "Ethnic food attitudes and behaviour among Belgians and Hispanics living in Belgium," *British Food Journal*, vol. 107, no. 11, pp. 823-840, 2005.

[6]   E. Paulson-Box and P. Williamson, "The development of the ethnic food market in the UK," *British Food Journal*, vol. 92, no. 2, pp. 10-15, 1990.

[7]   L. R. Marotz, Health, *Safety, and Nutrition for the Young Child*, Cengage Learning, Stamford, CT, USA, 2015.

[8]   S. M. Church, "EuroFIR synthesis report no 7: food composition explained," *British Nutrition Foundation Nutrition Bulletin*, vol. 34, pp. 250-272, 2009.

[9]   K. Argyri, E. Theophanidi, A. Kapna, C. Staikidoua, G. Pounisa, M. Komaitisa, C. Georgioub, and M. Kapsokefaloua, "Iron or zinc dialyzability obtained from a modified in vitro digestion procedure compare well with iron or zinc absorption from meals," *Food Chemistry*, vol. 127, pp. 716-721, 2011.

[10]   A. S. Adegoke, J. A. Akinyanju, and J. E. Olajide, "Fermentation of aflatoxin contaminated white dent maize (*Zea mays*)," *Research Journal of Medical Sciences*,

vol. 4, no. 3, pp. 111-115, 2010.

[11] M. Y. Monteiro, *Presença do Índio na Cultura Amazonense*, Edições Governo do Estado do Amazonas, Secretaria de Estado da Cultura, Turismo e Desporto, Manaus, AM, Brazil, 2001.

[12] A. G. Cunha, *Dicionário Etimológico Nova Fronteira da Língua Portuguesa*, Nova Fronteira, Rio de Janeiro, RJ, Brazil, 1991.

[13] R. C. Chisté, K. de O. Cohen, and S. S. Oliveira, "Study of tucupi physicochemical properties," *Ciência e Tecnologia de Alimentos*, vol. 27, no. 3, pp. 437-440, 2007.

CONCLUDING REMARKS

Fermentation has improved the nutritional and sensory qualities of foods as well as the food safety of human diets throughout the ages. Microorganisms are generous organisms whose simple enzymatic extraction of substrate energy results in more nutritive and palatable foods for humankind. The biotechnological advances have helped to solve limitations of food utilization: what is poor in one nutrient gets rich in it. There are certainly more improvements looming in the vast field of fermentation.

Printed by Books on Demand GmbH, Norderstedt / Germany